Separations Division 2014

Core Programming Area at the 2014 AIChE Spring Meeting and 10th Global Congress on Process Safety

New Orleans, Louisiana, USA
30 March - 3 April 2014

ISBN: 978-1-63439-079-8

Printed from e-media with permission by:

Curran Associates, Inc.
57 Morehouse Lane
Red Hook, NY 12571

CURRAN ASSOCIATES INC. proceedings.com

Some format issues inherent in the e-media version may also appear in this print version.

Printed by Curran Associates, Inc. (2014)

For permission requests, please contact AIChE
at the address below.

AIChE
3 Park Avenue
New York, NY 10016-5991

Phone: (203) 702-7660
Fax: (203) 775-5177

www.aiche.org

Additional copies of this publication are available from:

Curran Associates, Inc.
57 Morehouse Lane
Red Hook, NY 12571 USA
Phone: 845-758-0400
Fax: 845-758-2634
Email: curran@proceedings.com
Web: www.proceedings.com

TABLE OF CONTENTS

354012 Possible vs. Practical: Engineers Must Lead the Development of Practical Technologies

Abstract:

Chemical Engineering is need today more than ever. There is a strong societal desire for sources of energy and feedstocks that can sustainably support a high standard of living for all. Thermodynamic, material and economic challenges stand in the way. Most of the general public does not have the basic understanding of the energy flows, chemical transformations, and the scale of the systems that support our current way of life. These concepts are fundamental to chemical engineering. Chemical engineers should be on the vanguard of separating hype from truly practical options yet occationally we let society down. This talk will present several examples where the fundamentals of chemical engineering can be used to access the practicality of proposed solutions to energy and water issues, as well as explain why the venture capital model struggles in the materials market.

PROSYS™ Maximize Your Plant.

Operator Errors and What Can be Done to Minimize

Global Congress on Process Safety
March 31, 2014

10TH GLOBAL CONGRESS ON PROCESS SAFETY

PROSYS Maximize Your Plant.™

10TH GLOBAL CONGRESS ON PROCESS SAFETY

Presenter

Tom Nolan

- **Graduated from Ohio University with B.S. in Chemical Engineering**
- **24 years experience in chemical process industry in a variety of roles**
- **MAIC and DFSS Six Sigma Black Belt**

PROSYS Maximize Your Plant.™

10TH GLOBAL CONGRESS ON PROCESS SAFETY

The Cost of Errors

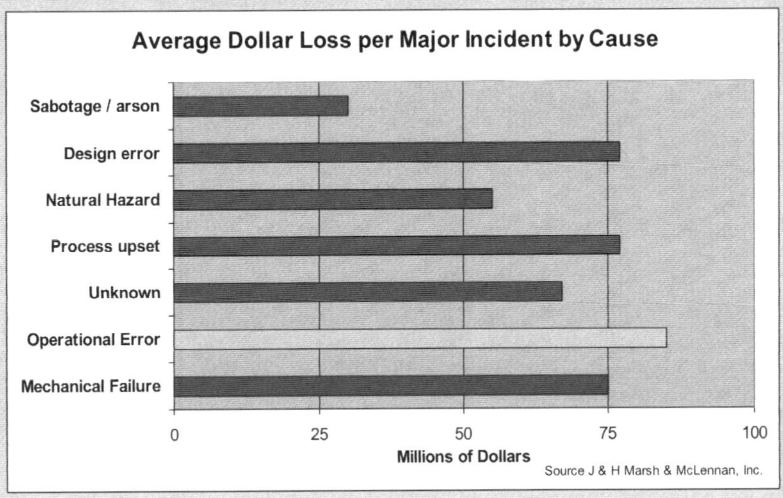

PROSYS Maximize Your Plant.™

10TH GLOBAL CONGRESS ON PROCESS SAFETY

The Cost of Operator Errors

- **ASM estimates total loss due to operator error is $8B per year**
- **Chemical Safety Topical Committee - average of one chemical incident per day - cost of over $2 million per incident to <u>comply with requirements</u>**
- **Errors cause 42% of unscheduled shutdowns**
- **70% of process incidents occur during start-up or shutdown**

PROSYS Maximize Your Plant.™

10TH GLOBAL CONGRESS ON PROCESS SAFETY

Reasons for Errors

- Lack of Skill
- Lack of Knowledge
- Carelessness
- System Design
  - Operator set up to make errors by inappropriate design – built in errors
  - Errors that are predictable are preventable by better design

PROSYS Maximize Your Plant.™

10TH GLOBAL CONGRESS ON PROCESS SAFETY

Focus of this Discussion

- System Design
  - **Alarm Management**
  - **Operator Graphics**
  - Operator set up to make errors by inappropriate design – built in errors
  - Errors that are predictable are preventable by better design

PROSYS Maximize Your Plant.™

10TH GLOBAL CONGRESS ON PROCESS SAFETY

Alarm Floods

In a number of industrial incidents, alarm floods were identified as a significant contributing cause to the incident...

As found by EEMUA in 1999 and CSB

Alarm Flood defined by ISA 18.2 as -
"10 or more annunciated alarms in any 10 minute period per operator"

PROSYS Maximize Your Plant.™

10TH GLOBAL CONGRESS ON PROCESS SAFETY

Why do Alarm Floods Occur?

One reason is not providing dynamic alarm management

- **Alarms need to indicate abnormal situations that require operator action**
- **Processes do not operate in one state**
- **What is normal vs abnormal changes with state**
- **Alarms are typically configured for run therefore many alarms are triggered upon a Change of State: Run to Shutdown**
- **Many of the alarms are not applicable or actionable for the new state impleading the operators ability to act quickly on what is important**

What is Impacted by Alarm Floods?

- **Product quality**
- **Operability or profitability of the process**
- **Loss of equipment**
- **Loss of containment – environmental releases**
- **Injury and loss of life in plant or community**

PROSYS™ Maximize Your Plant.™
10th GLOBAL CONGRESS on PROCESS SAFETY

What Makes Alarm Floods so Dangerous?

Can be a problem for three reasons:

- **A deluge of alarms can cause critical alarms to be missed**
- **Floods can be a significant distraction when dealing with process upsets**
- **Can be an indicator of larger systemic safety issues**

PROSYS Maximize Your Plant™

10TH GLOBAL CONGRESS ON PROCESS SAFETY

Impacting Alarm Management Design

- **Alarm rationalization is not a process to eliminate alarms - its about quality**
- **Good rationalization will add alarms when appropriate**
- **Typical (Static) rationalization is only for run mode**
- **Dynamic rationalization considers all plant modes**

PROSYS Maximize Your Plant™

10TH GLOBAL CONGRESS ON PROCESS SAFETY

Dynamic Alarm Management

- **Dynamic rationalization does everything a static rationalization does plus asks "when" for each alarm**
- **Requires dynamic software to make changes based upon operating mode of the plant**
- **Eliminates redundant alarms and lowers operator loading during transitions**
- **Only alarms what is abnormal and actionable for the given state**

PROSYS Maximize Your Plant.™

10TH GLOBAL CONGRESS ON PROCESS SAFETY

Actual Performance Metrics vs ISA 18.2

Type	Avg Alarm Rate/hr	Low/Hi of Avg Rate/hr	Peak Alarm Rate/hr	Low/Hi of Peak Rate/hr	% Time in Flood	Standing Alarms >24 hrs
Before Rationalization	30	Low=6.2 Hi=61	638	Low=152 Hi=2402	17.2%	9
ISA 18.2 Target Metrics	6	---	<60	---	<1%	<5
After Dynamic Alarm Management	2	Low=0.09 Hi=2.5	25	Low=13 Hi=42	0.25%	4

PROSYS Maximize Your Plant.™

10TH GLOBAL CONGRESS ON PROCESS SAFETY

Results After Dynamic Alarm Management

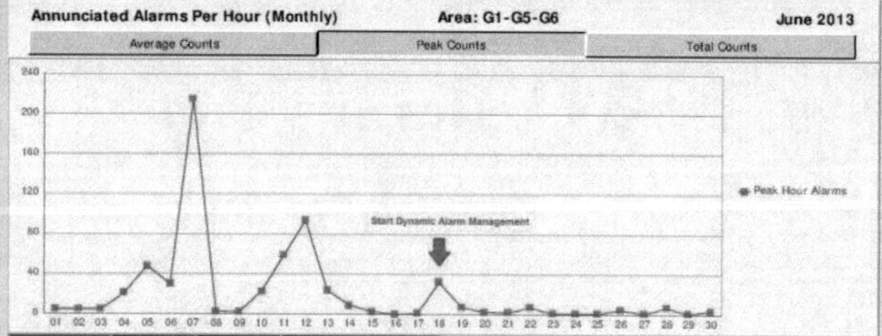

Day	Min Hour Alarms	Avg Hour Alarms	Peak Hour Alarms	Total Alarms
01-Jun-2013	0	0.3	5	6
02-Jun-2013	0	0.4	5	9
03-Jun-2013	0	0.5	6	13
04-Jun-2013	0	1.1	21	27
05-Jun-2013	0	13.3	48	319
06-Jun-2013	0	3.6	30	86
07-Jun-2013	0	11.9	215	285
08-Jun-2013	0	0.4	3	10
09-Jun-2013	0	0.3	2	7
10-Jun-2013	0	1.5	23	35
11-Jun-2013	0	4.7	59	112
12-Jun-2013	0	11.9	94	285
13-Jun-2013	0	3.8	24	90
14-Jun-2013	0	1.3	9	32
15-Jun-2013	0	0.3	3	8

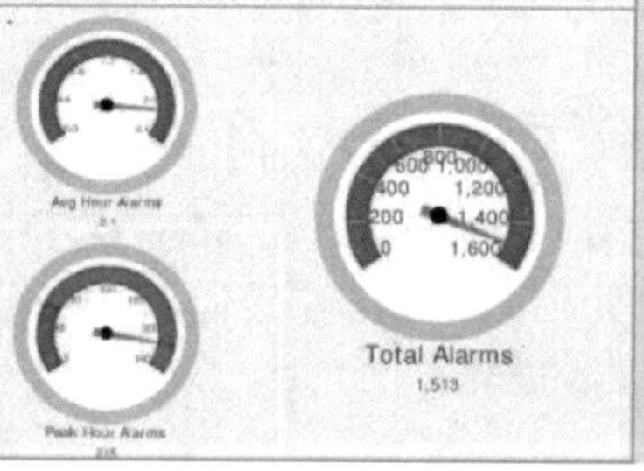

Operator Graphics – The Risk of Changing Focus

- Distraction of changing focus from process graphic to faceplate window
- Added workload from managing multiple open windows
- Increased probability of errors when changing values for un-intended tag with multiple faceplates open
- All of above become multiplied and more complex when process is transitioning from one state to another

PROSYS Maximize Your Plant™
10th GLOBAL CONGRESS ON PROCESS SAFETY

Faceplate Operation

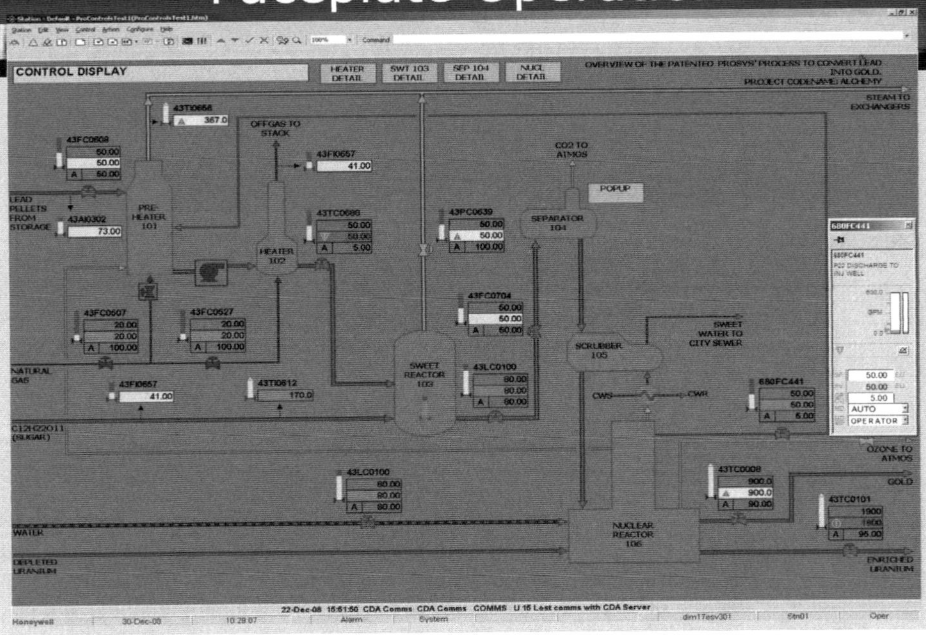

Potential for Errors

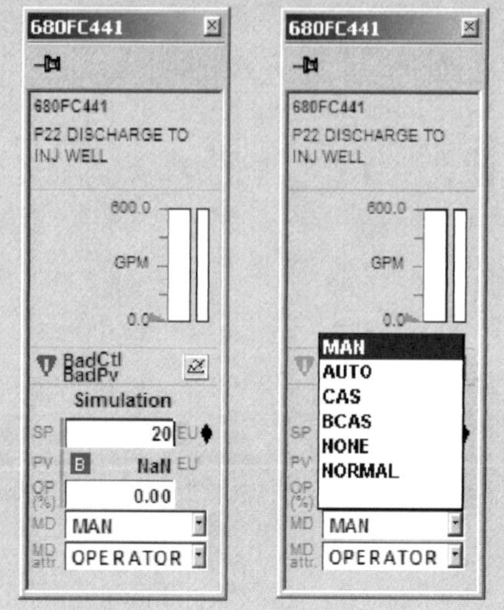

Faceplate Design -can introduce additional errors

- Mode drop down list covers SP, PV and OP values
- Operator may select wrong Mode if SP, PV and OP values are hidden from view
- Mode list offers more options than necessary

Direct Entry Operation

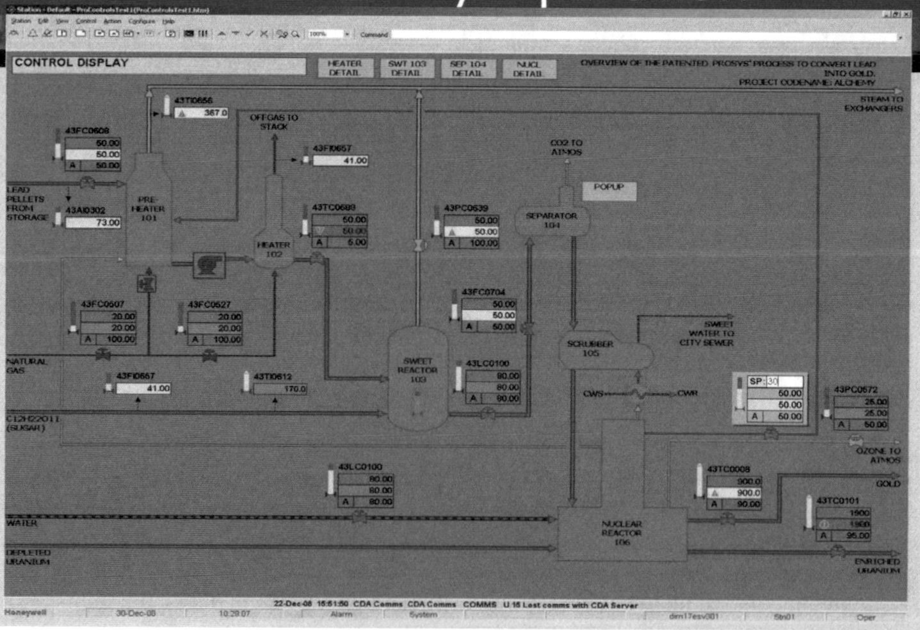

PROSYS Maximize Your Plant™

10th GLOBAL CONGRESS ON PROCESS SAFETY

Direct Entry Fields

Values can be entered by:

- Select point
- Typing value via keypad
- Pressing enter

Key Factors:

- By maintaining focus of points after value change, it improves safety for input errors, multiple clicking functions and is more time efficient
- Easily recognizable operator enterable fields

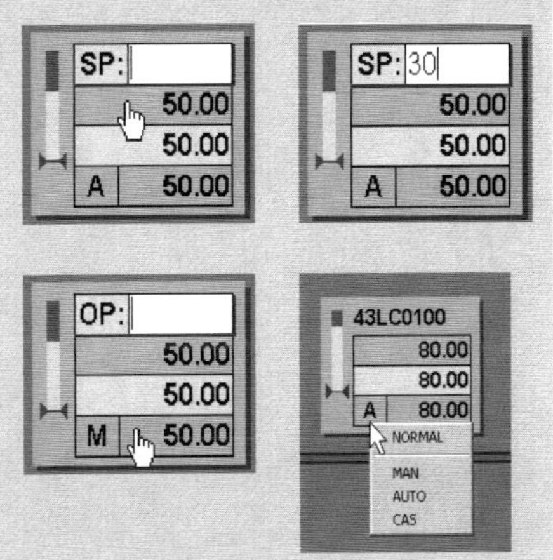

PROSYS Maximize Your Plant™

10th GLOBAL CONGRESS ON PROCESS SAFETY

Evaluation of Methods

Comparison of Faceplate to Direct Entry

- **Keystroke Level Analysis**
- **Operator Loading Analysis**
- **Risk Analysis**

PROSYS Maximize Your Plant™

10TH GLOBAL CONGRESS ON PROCESS SAFETY

Keystroke Level Model

Symbol	Time (s)	Description
K	0.28	Keystroke
P	1.1	Point to object
BB	0.2	Click on object
H	0.4	Home hands on keyboard or mouse
M	1.2+	Mental act or routine thinking

PROSYS Maximize Your Plant™

10TH GLOBAL CONGRESS ON PROCESS SAFETY

Estimate Execution Time Analysis

- **Evaluates execution time by an operator comparing faceplates to direct entry for SP/OP and Mode Changes**

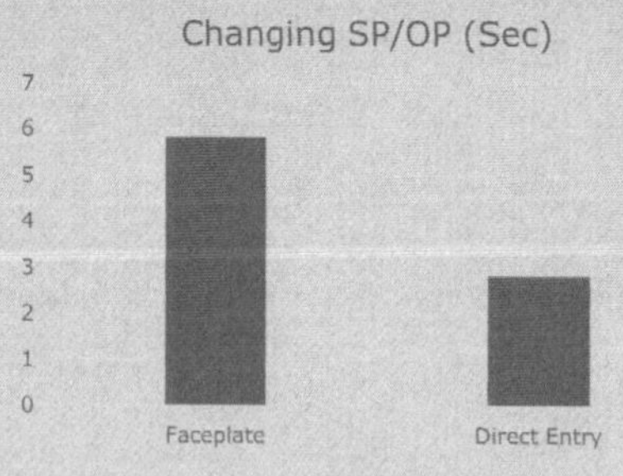

- **50 % Reduction**

PROSYS Maximize Your Plant™

10th GLOBAL CONGRESS ON PROCESS SAFETY

Estimated Operator Loading Analysis

- **Evaluates Mental and Physical Load on Operator**
  - Tasks that require very little thought are removed – such as (BB), (H)

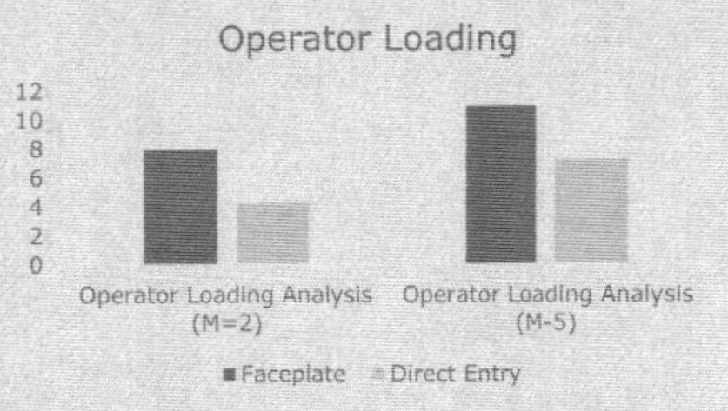

  - Mental and Physical tasks are weighted based on the amount of fatigue they cause – assigned a weight of M=2.0 to 5.0

PROSYS Maximize Your Plant™

10th GLOBAL CONGRESS ON PROCESS SAFETY

Risk Analysis

- **Identifies actions where operator entry errors can occur**

Risk Analysis

Faceplate operation KLM for changing SP/OP	Potential Error
Move to shape (P)	Select wrong point
Click on shape (BB)	
Move to faceplate (P)	Point in faceplate is previous point
Click on SP/OP field (BB)	Change wrong parameter
Move hands to keyboard (H)	
Type in value and press enter (4K)	Mistype value and press enter
Move hands to mouse (H)	
Move to close faceplate (P)	
Click to close faceplate (BB)	
Potential error count	**4**

PROSYS Maximize Your Plant.™

10TH GLOBAL CONGRESS ON PROCESS SAFETY

Risk Analysis

Direct Entry operation KLM for changing SP/OP	Potential Error
Move to SP/OP (P)	Select wrong parameter or point
Click on SP/OP field (BB)	
Move hands to keyboard (H)	
Type in value and press enter (4K)	Mistype value and press enter
Potential error count	**2**

PROSYS Maximize Your Plant.™

10™ GLOBAL CONGRESS ON PROCESS SAFETY

Direct Entry Vs Faceplate

Summary of Benefits

- **Execution Time SP/OP Changes - 51.5% less**
- **Execution Time for Mode Changes – 50% less**
- **Operator Loading – 33 % to 47% less**
- **Risk of Errors – 50% less**

PROSYS Maximize Your Plant.™

10™ GLOBAL CONGRESS ON PROCESS SAFETY

Conclusion

- **Many industrial errors are a result of operators using control systems with flawed designs**
  - Poor design and performance by alarm management distracts operators and/or occludes critical alarms
  - Operator graphics with multiple open faceplates can cause a change intended for one controller to be entered into the faceplate of another
- **The two mechanisms listed above are predictable and therefore preventable through better design available today**

PROSYS™ Maximize Your Plant.
Questions?
Comments?
10TH GLOBAL CONGRESS ON PROCESS SAFETY

Problem Statement

visium KMS
Process Safety Management Solution

- Any organisation can improve by leveraging knowledge and lessons learned.
- Knowledge can be externally or hidden locally, in a global enterprise, where lessons learned are difficult to share beyond their initial impact.
- There is so much data, information and knowledge (opportunities) coming at your company that can make you stronger or make you look like your not paying attention.
- Not paying attention or managing the available knowledge, can sink your ship… nobody wants that!

Expert Insight Delivered 2

OSyS
Part of the Rolls-Royce Group

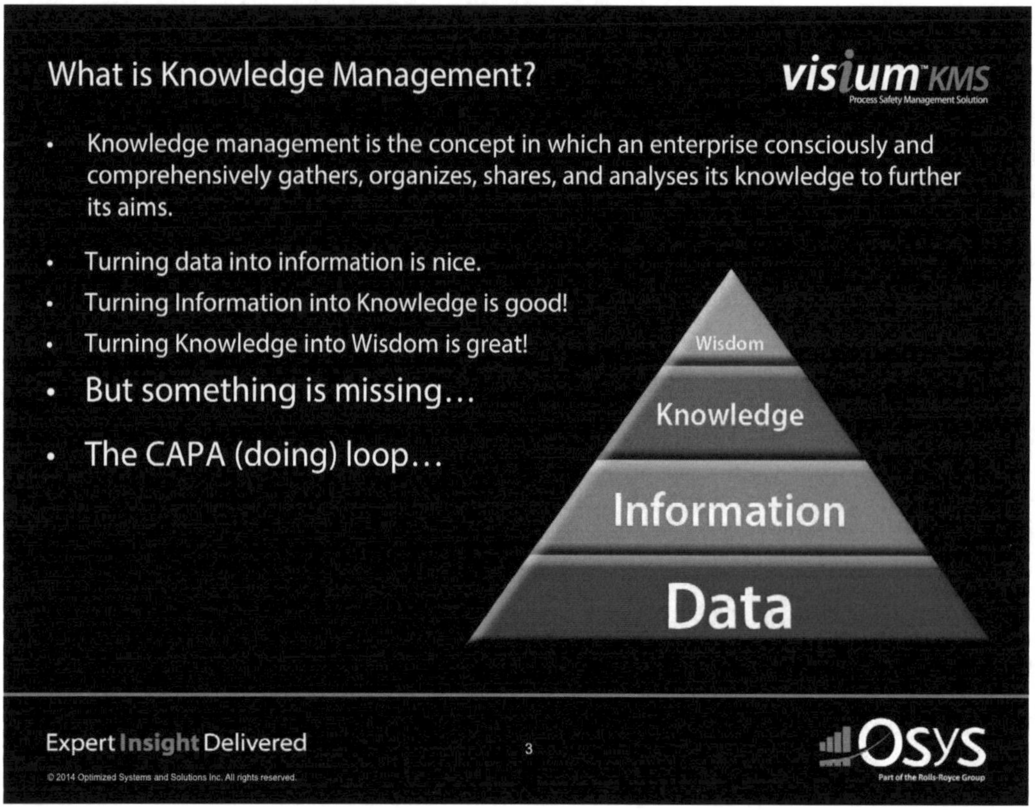
What is Knowledge Management?
visium KMS
Process Safety Management Solution
• Knowledge management is the concept in which an enterprise consciously and comprehensively gathers, organizes, shares, and analyses its knowledge to further its aims.
• Turning data into information is nice.
• Turning Information into Knowledge is good!
• Turning Knowledge into Wisdom is great!
• But something is missing…
• The CAPA (doing) loop…
Wisdom
Knowledge
Information
Data
Expert Insight Delivered
3
© 2014 Optimized Systems and Solutions Inc. All rights reserved.
Osys
Part of the Rolls-Royce Group

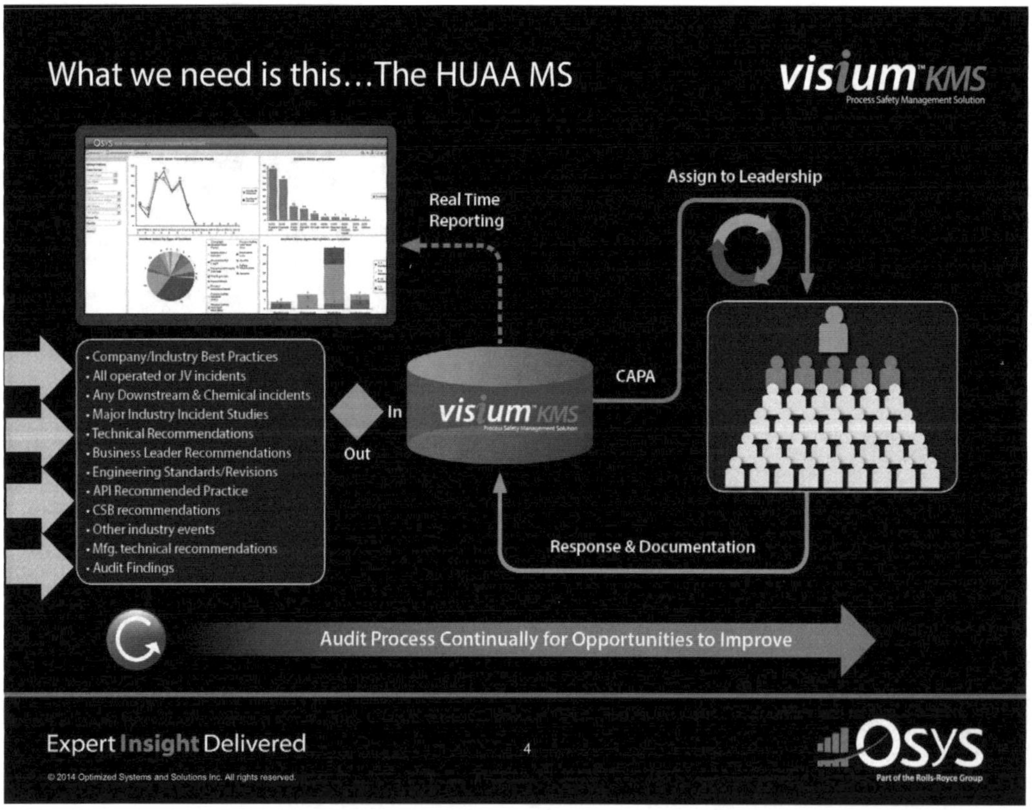
What we need is this…The HUAA MS
visium KMS
Process Safety Management Solution
Real Time Reporting
Assign to Leadership
• Company/Industry Best Practices
• All operated or JV incidents
• Any Downstream & Chemical incidents
• Major Industry Incident Studies
• Technical Recommendations
• Business Leader Recommendations
• Engineering Standards/Revisions
• API Recommended Practice
• CSB recommendations
• Other industry events
• Mfg. technical recommendations
• Audit Findings
In
Out
visium KMS
CAPA
Response & Documentation
Audit Process Continually for Opportunities to Improve
Expert Insight Delivered
4
© 2014 Optimized Systems and Solutions Inc. All rights reserved.
Osys
Part of the Rolls-Royce Group

What is the HUAA Management System…

visium KMS
Process Safety Management Solution

HUAA – Heard, Understood, Acknowledged, Actioned.

1. Opportunities identified (heard) by your experts,
2. Captured by an electronic system,
3. Expertly Reviewed (understood),
4. Accepted or rejected (acknowledged) and
5. Assigned (actioned) to Sr. Managers for resolution and tracked to closure.

Expert Insight Delivered 5

OSyS
Part of the Rolls-Royce Group

HUAA – Management Systems - Detail

visium KMS
Process Safety Management Solution

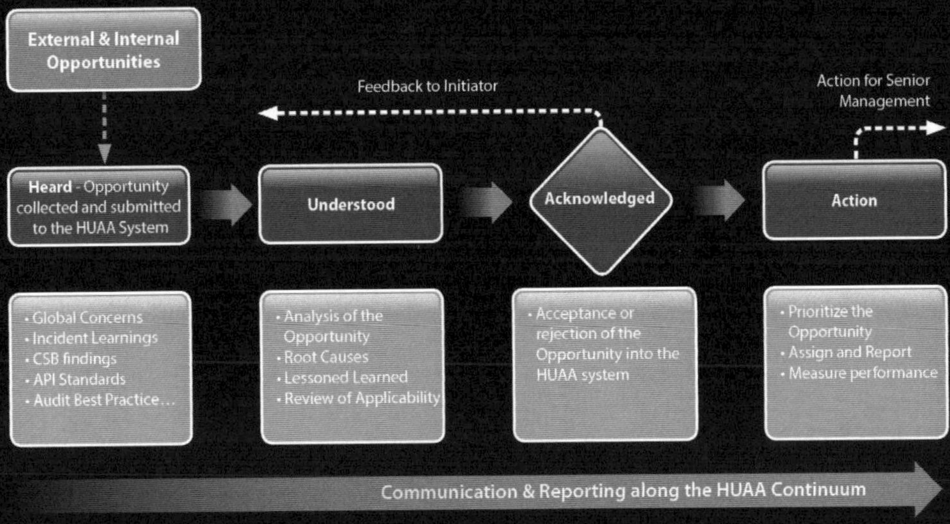

Expert Insight Delivered 6

OSyS
Part of the Rolls-Royce Group

HUAA Results

visium™ KMS
Process Safety Management Solution

- Efficient Collection - Important Opportunities are identified and collected from many sources (internal and external), and entered into the HUAA management system.
- Expert Vetting - These opportunities are vetted, parsed and resolved for inclusion or exclusion; documenting & communicating both.
- Accountable Assignment - Included opportunities are systematically assigned to Sr. Leaders and managed across the enterprise at SVP and C levels.
- Action - Improvement will be assured through the accountable actions of these leaders.
- Results - The organization will gain competitive advantage by active listening, learning from their experience, as well as the experience of others, and by taking ACTION!

Expert Insight Delivered 7

OSyS
Part of the Rolls-Royce Group

Sharing is not enough to keep your ship afloat!

visium™ KMS
Process Safety Management Solution

Expert Insight Delivered 8

OSyS
Part of the Rolls-Royce Group

352631 Autocad - Smart Grid – Agriculture - Architecture and the United States Government

Abstract:

The computational methods, process control, process design, process safety, public policy of the smart chip for agriculture industry, and architecture industry in the United States utilizing the computational methods of AutoCAD in the structural development stages of the smart grid for device/component/equipment and geographical development and public policy for grant funding.

352747 Production, Characterization and Catalytic Studies of Biobased Carbon Materials

Abstract:

Carbon-based nano materials were produced from wood chips, lignin and bio-char. The influence of reaction conditions on the production of nano-carbons was investigated. Carbon-based nanomaterials were characterized using scanning electron microscopy (SEM), transmission electron microscopy (TEM), high resolution transmission electron microscopy (HRTEM), and X-ray diffraction (XRD). The catalytic performance of these nanomaterials as syngas cleaning adsorbents and chemical conversion catalysts for bio-fuel processes has been evaluated.

347835 Uncertainty in Sour Gas Viscosities Estimation, What Is the Effect on Your Reservoir Inflow and Tubing Performance

Abstract:

In light of the need to find an applicable, consistent and promising correlation to accurately calculating sour gases viscosity under several reservoir conditions, this study aims to compare the existing correlations and other industry standard methods, check their accuracy and range of applicability for sour/acidic gases. Moreover, investigating the impact of uncertainties in estimating sour gas viscosities on reservoir flow performance relationship (IPR), reserves and resources estimations, tubing performance, pipeline and facilities design, the design of compressors and more importantly the need to find such correlation to be used in further advancements of reservoir simulation.

354052 Chromonic Nanocarriers for Chemotherapeutics: Size Distribution and Control Release Studies

Abstract:

Conventional chemotherapy fails to release the drug in the controlled manner; rather they release the drug instantly due to which re-administration is required as therapeutic index falls. Moreover, they target both the cancer cells and non-cancer cells which make the treatment of the cancer cells highly ineffective due to excessive toxicities. Recently, the nanoparticle mediated drug delivery for chemotherapeutics has gained much attention. The present study reports the use of a novel class of materials – *Chromonics* in cancer drug delivery. Folic acid as a chromonic material has been explored to develop nanocarrier for sustained release of chemotherapeutic drugs *in-vitro*. With the help of a biocompatible polymer as a tool, folate nano-domains were formed from self-assembled liquid crystalline folate solutions. These nano-domains were crosslinked to form a stable nanoparticle system. The effect of different factors like crosslinking agent, relative polymer and folate concentration, on the size distribution was investigated in detail. Based on the observations, we have reported that not only the synthesis but also the size of the developed nanoparticles can be controlled by varying the folate: HPMC weight ratios. The present study then explored the release of drugs from these nanocarriers. We have investigated the release of folic acid as a model drug from these nanoparticles as well as release of methotrexate and cytarabine as an anticancer drug for the period of 60 days. It was observed that release of drug was nanoparticle's size-dependent, release medium dependent, pH of the release medium dependent, crosslinking agent and its concentration dependent. The significant part of the current study is the synthesis of chromonic nanoparticles, how the size can be controlled, how controlled release can be obtained as well as how the release profiles can be modified depending on the dosage forms. Therefore, we can say that folate nanoparticles can be a potent nanocarrier for the sustained release of cancer drugs.

Double Containment Piping Systems: A Fail-Safe, Leak-Free Solution

IPEX
Committed to Excellence

Darin Johnson, P.Eng.
Applications Engineer
IPEX USA LLC

AIChE – 2014 Spring Meeting
New Orleans, LA

Monday, March 31, 2014

Why Double Containment Piping Systems?

IPEX
Committed to Excellence

- December 1988 – law change within the ***Resource, Conversation, and Recovery Act*** in the United States regarding underground piping.
- U.S. Environmental Protection Agency (EPA) requirement: Code of Federal Regulations (CFR), Title 40, Part 280.
  - Petroleum Products
  - Hazardous Substances (over 3200, CFR Title 40 Part 302.4)
  - Hazardous Wastes

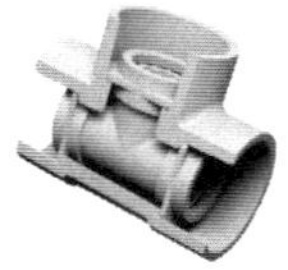

Why Double Containment Piping Systems?

- Buried tank - all buried piping connected to it must be double contained.
- Above ground tank – associated buried piping consists of 10% or more of the tank volume, it also must be double contained.

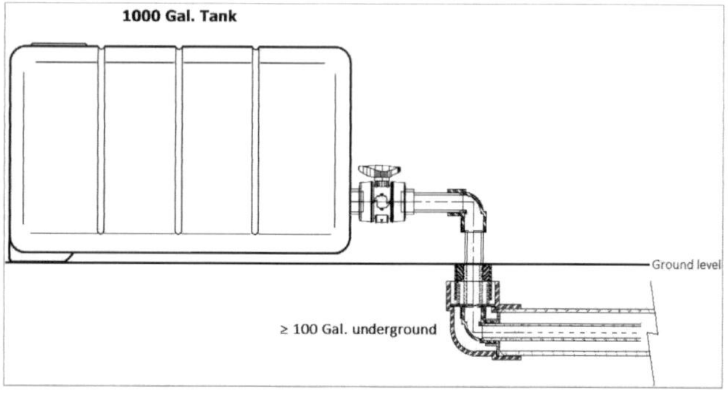

Why Double Containment Piping Systems?

- If the piping is less than 10% of the volume of the aboveground tank, and you choose not to double contain the buried piping, ***responsibility for any clean-up or remediation resulting from a release will still fall on the Client.***

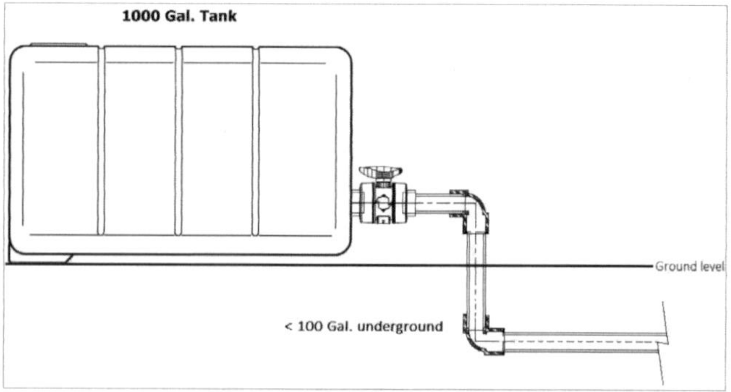

Why Double Containment Piping Systems?

- **Above Ground Piping Systems**
  - Different jurisdictions have to ability to require chemicals be double contained.

Nomenclature of Double Containment Piping

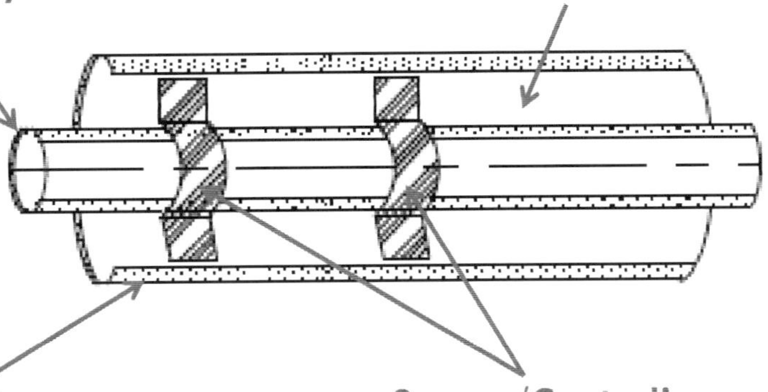

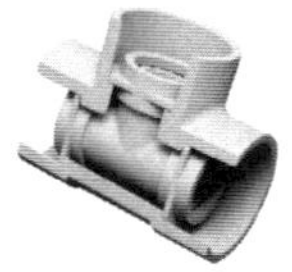

IPEX
Committed to Excellence

Double Containment Piping Systems: Material Combinations

SYSTEM Carrier	Containment	Carrier SIZE inches
PVC 80	PVC 80	1/2 - 12
PVC 80	PVC 40	1/2 - 12
PVC 40	PVC 40	1/2 - 12
PVC DWV	PVC DWV	1-1/2 - 12
PVC 80	CS SCH 10	1/2 - 12
CPVC 80	CPVC 80	1/2 - 8
CPVC 80	PVC 80	1/2 - 12
CPVC 80	PVC 40	1/2 - 12
CPVC 80	CS SCH 10	1/2 - 12
NAT PP 80	PVC 80	1/2 - 2
NAT PP 80	PVC 40	1/2 - 2
RED PVDF 80	CPVC 80	1/2 - 2
RED PVDF 80	PVC 80	1/2 - 2
NAT PVDF 80	PVC 80	1/2 - 2
ENFUSION PP	PVC DWV	1/2 - 6
S40 A53B CS BW	PVC 80	1/2 - 8
S40 A106B CS BW	PVC 80	1/2 - 8

SYSTEM Carrier	Containment	Carrier SIZE inches
S40 A53 BLK SW	FRP	1/2 - 2
S40 A53 BLK SW	PVC 80	1/2 - 2
S80 A53B CS SW	PVC 80	1/2 - 2
S10 T304L SS BW	PVC 80	1/2 - 8
S10 T304L SS BW	PVC 40	1/2 - 8
S10 T0304L SS BW	CPVC 80	1/2 - 8
S10 T304L SS BW	FRP	1/2 - 12
S10 T304L SS BW	S10 T304L	1/2 - 12
S40 T316L SS BW	PVC 80	1/2 - 8
S40 T316L SS BW	CPVC 80	1/2 - 8
S40 T316L SS BW	FRP	1/2 - 12
S40 T304L SS SW	PVC 80	1/2 - 2
S40 T304L SS SW	PVC 40	1/2 - 2
S40 T304L SS SW	CPVC 80	1/2 - 2
S40 T304L SS SW	FRP	1/2 - 2
S10 T316L SS SW	PVC 80	1/2 - 2
S10 T316L SS SW	CPVC 80	1/2 - 2

IPEX
Committed to Excellence

Double Containment Piping Systems: Design Considerations

- **Material Selection**
  - Primary Pipe
    - Chemical Compatibility
    - Pressure Capabilities
    - Thermal Properties
  - Secondary Pipe
    - Chemical Compatibility
    - Loading (Static & Live) for buried applications
    - Hanging requirements for above ground
    - Leak detection requirements
    - UV requirements

Double Containment Piping Systems: Design Considerations

- **Thermal Expansion & Contraction**
  - Two basic thermal relationships may be present:
    - When only one of the two systems experiences thermal expansion/contraction.
    - When both systems experience different magnitudes of thermal expansion/contraction.
  - The Floating System
  - The Restrained System

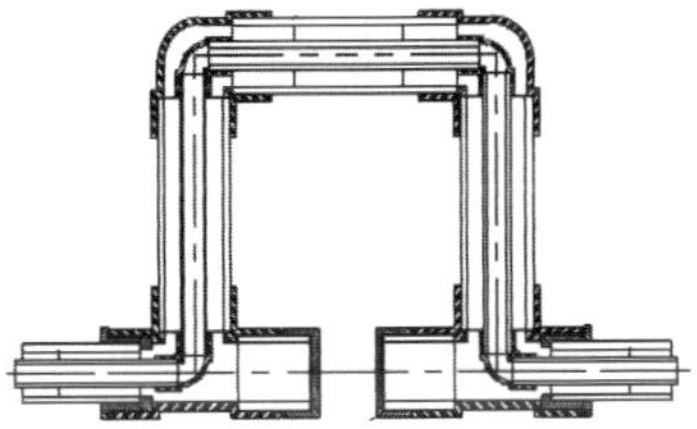

Double Containment Piping Systems: Leak Detection Requirements

- EPA requirement: (CFR), Title 40, Part 280
  - Pressurized systems require electronic leak detection that must be checked once per year.
  - Drainage systems (gravity flow) require visual leak detection with monthly manual inspection for a media leak.

Double Containment Piping Systems: Leak Detection Requirements

- The following forms of leak detection are acceptable as per regulations
  - Visual (drainage applications only)
  - Electronic low point probes
  - Continuous cable

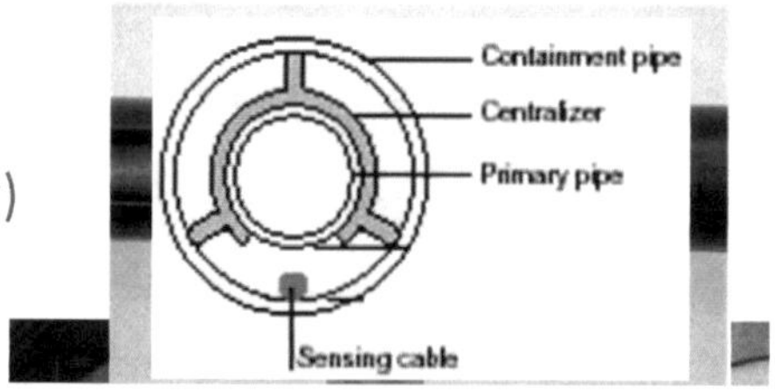

Double Containment Piping Systems: Applications

- Underground Applications
- WTP & WWTP
- Oil & Gas
- Hospitals
- Museums
- Food and Beverage
- Pharmaceuticals
- Laboratories
- Data Centers

IPEX
Committed to Excellence

Double Containment Piping Systems: A Fail-Safe, Leak-Free Solution

Questions?

Thank you!

DTP Process

"Now Ready to License"

JGC Corporation
Mitsuo Morita

2014

JGC CORPORATION

MITSUBISHI CHEMICAL

What is DTP Process ?

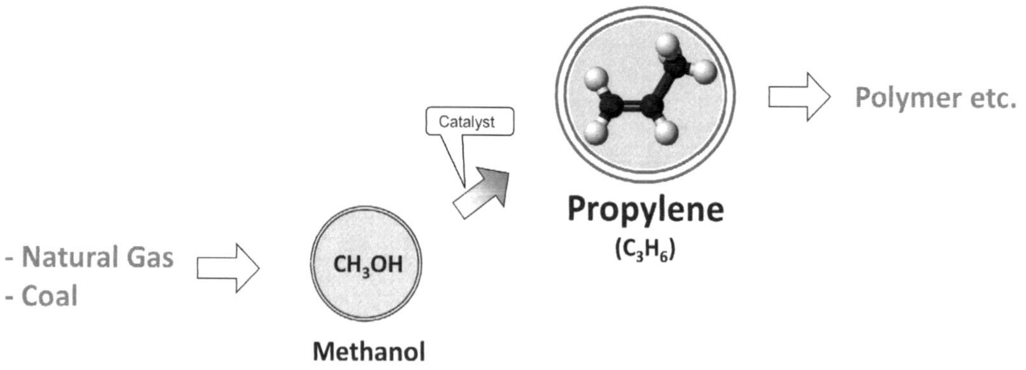

-On-Purpose Propylene Production from Methanol

JGC CORPORATION

- 2 -

MITSUBISHI CHEMICAL

What is DTP Process ?

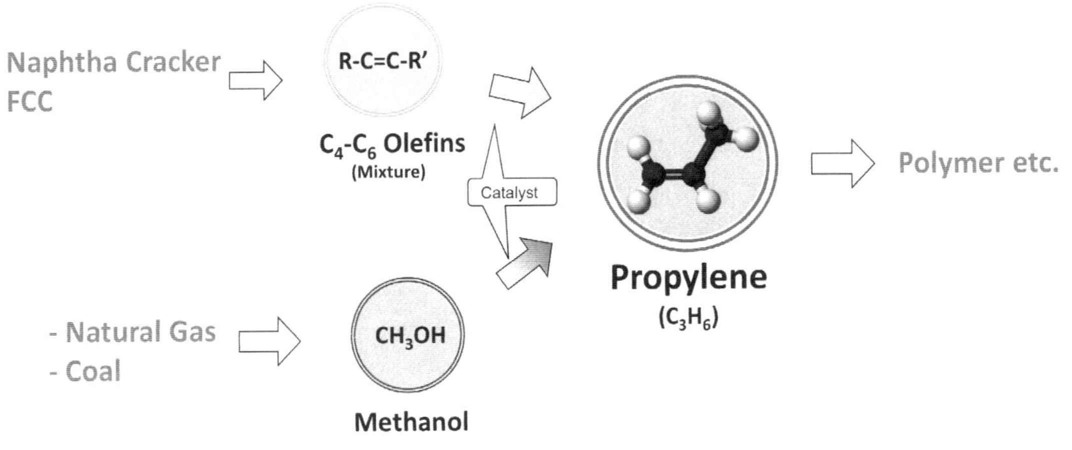

-On-Purpose Propylene Production from Methanol

- Additional Unutilized C_4-C_6 Olefins as Feedstock

Development History and Prospect

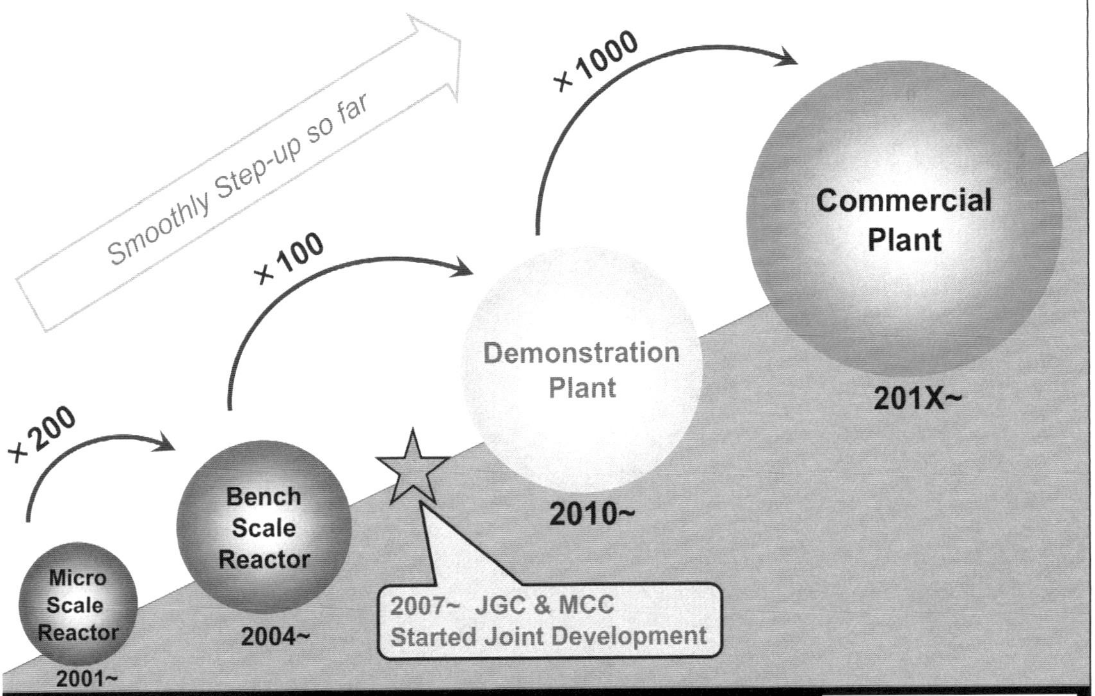

Demonstration Plant Overview

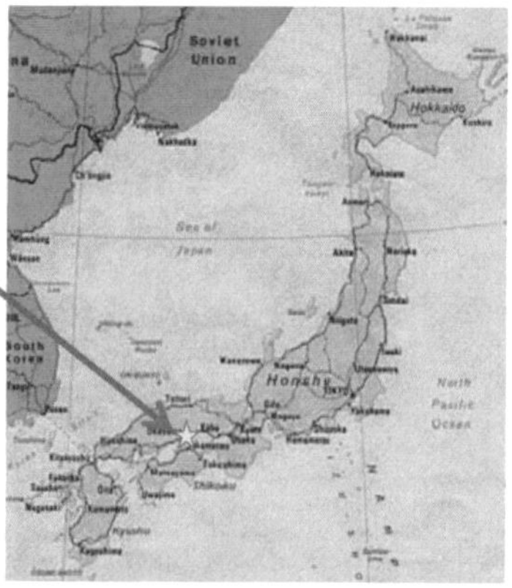

- Located in Mitsubishi Chemical Mizushima Plant
- Successfully Start-up in August 2010
- Continuous Operation for Two Years

Process Flow (Standalone Case)

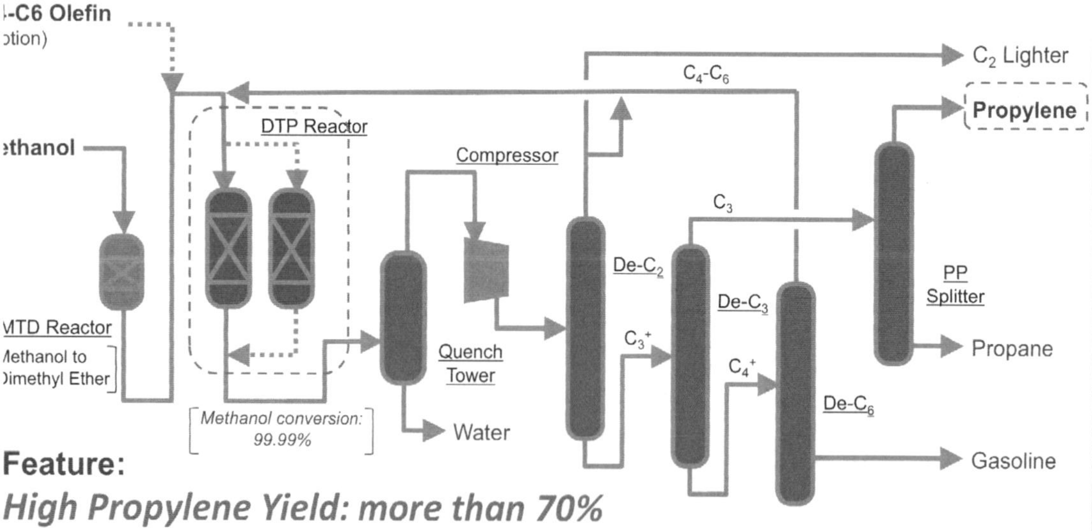

Feature:

High Propylene Yield: more than 70%

Applied Existing Proven Technologies

Process Flow (Integration Case)

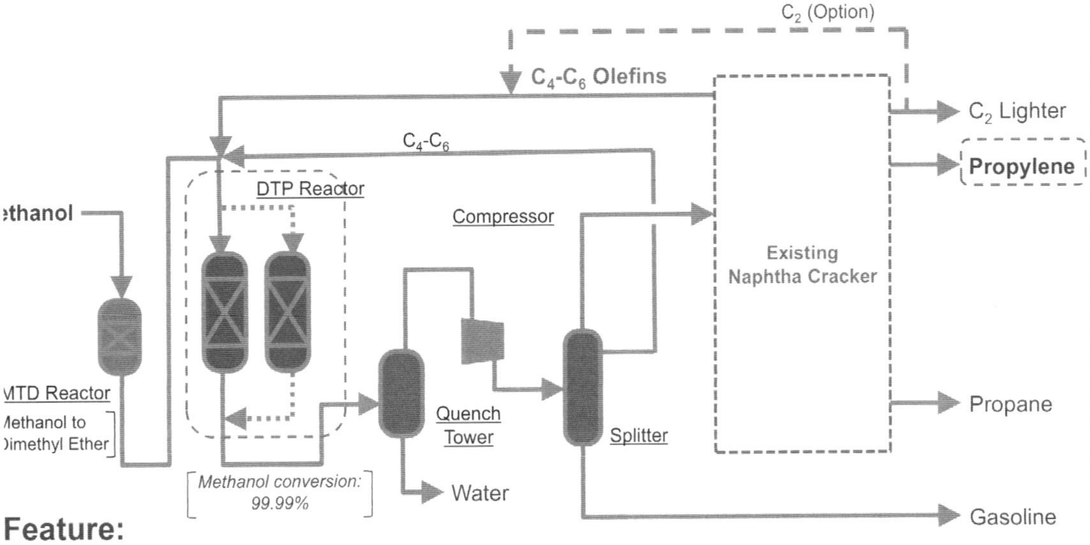

Feature:

High Propylene Yield: more than 70%

Applied Existing Proven Technologies

Low Investment Cost in case of Integration with Naphtha Cracker

Application

1) Naphtha Cracker Integration Case (Propylene Capacity: 20KTA – 200KTA)

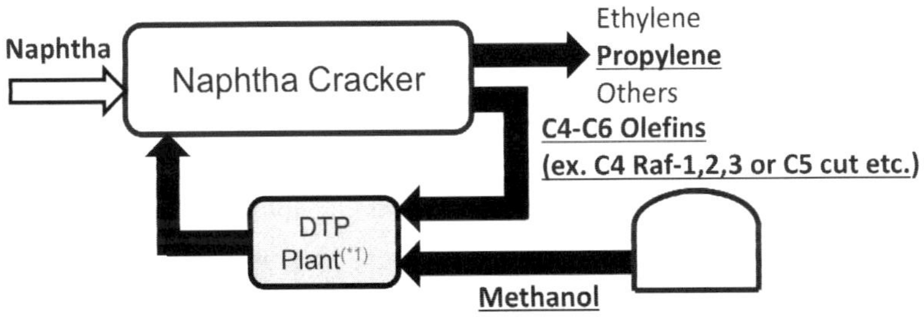

(*1) Utilizing facilities in Naphtha Cracker.

2) Standalone Case (Propylene Capacity: 250KTA – 500KTA)

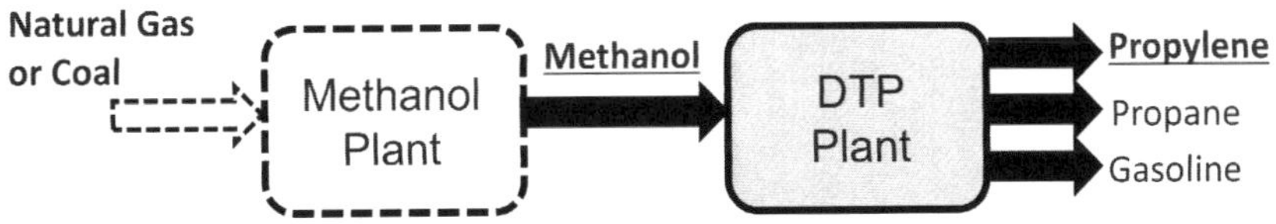

- Two Cases available depending on Client requirements !!

DTP is…

- *High Yield Propylene Production Process*
- *Flexible Feed Methanol to Olefin Process*
- *Process that increase P/E Ratio for Cracker*

Thank you very much !!

Contact:

Mitsuo Morita
Market Development Dept.
Global Marketing Division
2-3-1, Minato Mirai, Nishi-ku,
Yokohama 220-6001, Japan

Phone: 81-45-682-8371
FAX: 81-45-682-8183
E-Mail: morita.mitsuo@jgc.co.jp

352533 Self-Cleaning "Bernoulli" Type Filters Used in Onshore and Offshore Applications

Abstract:

Georg Schuenemann GmbH (SAB-BREMEN) has developed a new Filter Type F-480, dubbed the "Compact Giant" for applications in onshore water intake and cooling water systems, sea water desalination plants, and in offshore projects, such as FPSO's, Drillships, and Production Platforms.

This paper will describe the essential design features of the filters, based on the "Bernoulli principle" which result in a fully automated self-cleaning filter system, and choice of materials in a sea water service with high corrosion rates.

A case study will be presented in which stringent project requirements such as substantial flow rates, adaptibility to space limitations and pipe connections had to be observed. Reliability of the system, extended run times without the need for frequent replacement of filter elements, as well as ease of monitoring and operator involvement have been achieved.

352653 Effects of Support on Sulfur Tolerance and Regeneration of Pt Catalysts Measured By Ethylene Hydrogenation and EXAFS

Abstract:

We investigated the effect of support on sulfur tolerance and regenerability under reducing environments by measuring the rate of ethylene hydrogenation, combined with hydrogen chemisorption, and extended x-ray absorption fine structure (EXAFS). Alumina and titania (P25) supports were used to prepare Pt catalysts at 1 wt.%. The catalysts were tested after treatment in H_2S/H_2 at 250 °C followed by regeneration treatments in H_2 at 250 °C, 350 °C and 450 °C. The sulfided catalysts exhibited a 20-27 times decrease in the rate of ethylene hydrogenation and a 4-6 times drop in the Pt surface area. Regenerations up to 450 °C were unable to remove all the sulfur, as shown by the presence of Pt-S bonds by EXAFS. However, a partial recovery of the hydrogenation rate per mole of Pt was observed on sulfided Pt/Al_2O_3 after reduction at 450 °C, while the induction of strong metal support interactions (SMSI) at reduction temperature above 350 °C was observed on Pt/P25, regardless of the presence of sulfur. For Pt/P25, the reversal of the SMSI state together with sulfur removal by mild oxidation suggests that sequential reduction/oxidation treatments may be more effective in restoring the S-free state of TiO_2-supported catalysts.

352799 Improved Operational Efficiency and Reliability through Insulation Materials Selection

Abstract:

This paper will explore the implications of insulation materials selection on the energy efficiency and reliability for Ethylene production units. Considerations will include insulation properties, structural support, installation methods, installation and maintenance costs.

BOILER MACT COMPLIANCE: MERCURY IN FUEL GAS

PATRICK LAINE, WES YONGE, & VIC VICKERY
PEI, HOUSTON, TX 77043

Patrick Laine, Ph.D.
Chief Scientist
Houston, TX
850-866-6686
plaine@pei-tx.com

Wes Yonge, PE
Chief Engineer
Houston, TX
713-806-8396
wyonge@pei-tx.com

James 'Vic' Vickery
Chief Technology Officer
Houston, TX
281-536-0899
vvickery@pei-tx.com

1 Introduction & Objectives

For most ethylene producers the primary objective of this sampling and analysis effort will be to verify and "establish a mercury fuel specification through which gas-fired boilers that use a fuel other than natural gas or refinery gas may be considered Gas 1 unit" as identified in 40 CFR Part 63.7521.(f).

Measurement and Monitoring Requirements: Decision Rules

- ❑If Hg concentration ≤ 50% of Gas 1 specification* then no further testing required
- ❑If Hg concentration > 50% and ≤ 75% of the specification* then semi-annual sampling required
- ❑If Hg concentration > 75% of the specification* then monthly sampling required

*Specification is **40 µg/Sm³**

The goal will be to execute a sampling and analysis campaign that will produce compliance quality data. The key to producing and validating compliance quality data is a well-designed and executed QA/QC protocol. At a minimum, a performance based QA/QC protocol is recommended that includes the following components:

- Sorbent Trap Breakthrough Bed:
  - ❑ Sorbent traps consisting of a minimum of two sorbent beds in series so that the trailing bed can be analyzed to verify that no breakthrough occurred from the leading bed.
- Duplicate Samples:
  - ❑ The sampling system should be configured with two trains to accommodate the collection of simultaneous duplicate sorbent trap samples, in parallel, providing a means of verifying precision in the sample collection process.
- NIST (National Institute of Standards and Technology) Traceable Spike:
  - ❑ Sorbent traps with a third bed (downstream) that is spiked with a NIST traceable spiked mass of mercury serves as a matrix spike and provides a means of verifying that components in the matrix (the fuel gas) are not affecting the trapping efficiency of the sorbent trap.

2 Methods

Sorbent Trap Mercury Sampling Methods Comparison

Method	Description	Method Type: S = Sampling A = Analysis	Evaluation	Sample Collection Phase QA/QC
ISO 6978	Hg in Natural Gas by Thermal Desorption (CVAFS)	S & A	Method calls for low sample volumes can lead to non representative results.	No
EPA Method 30B	Determination of Gaseous Hg Emissions from Coal Fired Combustion Sources using Carbon Sorbent Traps	S	Requires modification (Modified M30B)	Yes
ASTM D6350	Hg S&A in Natural Gas (AFS)	S & A	Method calls for short term samples (i.e., low sample volume) which leads to poor accuracy.	No
ASTM D5954	Hg S&A in Natural Gas (AAS)	S & A	AAS is less sensitive than AFS.	No
EPA Method 1631	Measurement of Hg in Water	A	No	Yes

3 EPA Method 30B QA/QC Performance Criteria

QA/QC Test or Specification	Acceptance Criteria	Frequency
Analyzer calibration	Correlation Coefficient: 0.999	Daily
Transfer efficiency of analyzer	%Transfer Efficiency: 90%–110%	Once per sorbent trap
Initial calibration blank	< 0.5 ng	Once before each calibration
Sorbent Trap Section B breakthrough	≤ 10% of Section A Hg mass, for Hg concentrations > 1µg/dscm ≤ 20% of Section A Hg mass, for Hg concentrations ≤ 1µg/dscm	At each sample point
Spike Recovery Test	Average recovery between 75% and 125 % for Hg_0	Once per sample point
Paired Sorbent Traps Agreement	≤ 10% RD for Hg concentration > 1µg/dscm ≤ 20% RD or 0.2 µg/dscm absolute difference for Hg concentrations ≤ 1µg/dscm	At each sample point

4 Equipment

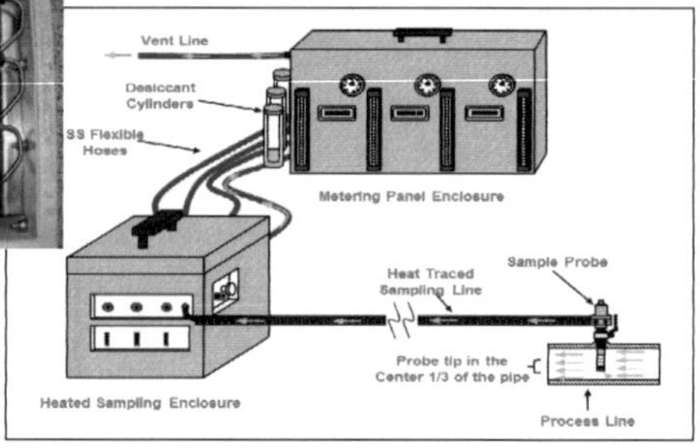

Flow →	Primary Bed	Breakthrough Bed	Spiked Bed
QA/QC Acceptance Criteria:	99% of Total Collected Hg	≤ 10% of Primary Mass	Recovery between 75% and 125%

Gold Trap:

5 Conclusions/PEI Recomendations

- ❑ There are five methods accepted by the Boiler MACT rule. PEI has extensive experience and expertise employing **all five** of these methods.
- ❑ Operators can achieve compliance without submitting a Fuel Analysis Plan by using ASTM D6350, however, we have noticed variable results with this method.
- ❑ The **PEI recommendation** for obtaining the most accurate compliance quality data is to use Modified EPA Method 30B in conjunction with EPA 1631 where traps should be analyzed in a certified ISO 17025 laboratory. The recommended pathway includes a rigorous, performance based QA/QC protocol that validates accuracy and precision of all sample collection procedures.

351376 Young Professional Simulation Tutorial

Abstract:

18C00 Young Professional Simulation Tutorial

Authors: Naomi Hua and Mike Donahue, Invensys

Process simulation is a complex skill to master. Even with years of experience, a typical user is constantly facing new problems that need to be addressed. There are many issues that can lead to unexpected and unnecessary errors with the use of process simulation tools. Regardless of what simulation program you are using, these issues are common from one program to another. In this presentation, targeted for the young professional, we will discuss the "Most Common Pitfalls in Process Simulation". For each topic, we will discuss various techniques and methods these troublesome problems and prevent their occurrence.

Topics to be discussed include:

1. Erroneous data entry
2. Simulation defaults
3. Bad plant data
4. Thermodynamic method selection

5. Simulation tolerances and convergence

6. Recycle identification and handling

7. Model complexity

8. Difficult or impossible specifications

9. Assay pseudo-component development

10. Assay property blending

1. Erroneous data entry is the most common mistake made when building a process simulation model. This presentation will cover methods that can help ensure proper data entry in your process simulation model, such as checking the data you entered, particularly the units of measure, using "alternate forms" of data input such as copy/paste and data import, and reviewing all warning and error messages.

2. Defaults are methods, values, and conditions that are automatically selected for you by the program when you begin a new simulation model. These defaults will affect the results calculated by your model and every simulation program has them. Most defaults within simulation programs fall under categories such as convergence criteria, thermodynamic methods, tolerance and units of measure. The problem with defaults is unintended results from not fully understanding the effects a particular default has on the model. Take the time to understand the defaults of the program and how these defaults will affect the results. If you do not agree with the defaults selected by the program, most simulation programs will allow the user to change the default.

3. Plant data changes over time. The feed information and process conditions of the unit operations may be different. Not only does plant data change over time, but measurement information is often very inaccurate. This could be due to faulty or poorly calibrated measurement equipment, leaks in lines, sensor failures, and other issues. Because of these changes and inaccuracies over time, you always have to validate the information that is entered in your model to ensure that you are obtaining the proper results. To do so, you can utilize programs that can help with data validation and reconciliation like SimSci's ROMeo online optimization program. With it, you can connect your model to your data historian to constantly import current data measurements into your model and then reconcile these measurements to determine when problems arise within your plant and where those problems are occurring. You will also need to use your engineering judgment to determine if the data is valid. More details will be provided during the presentation.

4. The proper selection of thermodynamic methods for your simulation is the most important decision that you make when modeling a process. The method you select could have a drastic effect upon your results. Often, any number of thermodynamic methods is suitable for a given application. It is up to the user to determine the best method to use based upon knowledge, personal experience, and best judgment. Some important factors that need to be considered are the components being used and the operating conditions.

Oftentimes, the selected thermodynamic method still needs to have additional inputs such as binary interaction parameters to simulate the process accurately.

5. Many of the calculations in process simulation are iterative. When these calculations iterate, the program compares the results calculated from one iteration to the results calculated in the next iteration in order to determine

whether it is converged to a solution or needs to continue calculating. This method of comparing results is where tolerances come into play and simulation programs use tolerances to determine when a solution is reached. One of the major mistakes when modeling a process is not understanding or using the correct tolerance values. A looser tolerance may make the file easier and quicker to converge, but it also reduces the accuracy of the model. A tighter tolerance may be more accurate, but might make it impossible for the model to solve or increase the convergence time exponentially. A common practice is that when you have reached a converged solution with your model, tighten the tolerances within the model and solve it again. If you notice that your answers have changed significantly, you need to continue tightening the tolerance values until your results change little from one iteration to another. You will then know that you have reached an accurate solution for your model.

6. Recycles in process simulation occur when you return to a previously calculated unit operation to recalculate the unit due to a change made elsewhere in the model. They are very common and can often be difficult to converge. Some suggestions on troubleshooting recycle convergence include initially solving your model without the recycles to see how it behaves, providing initial estimates for recycle streams, using tighter tolerances on unit operations within recycle loops than the total recycle tolerances themselves, and ensuring that all components have a way in and out of a recycle by performing mass balances and using purge or makeup streams where necessary.

7. The general rule of thumb when modeling is to start simple and add complexity as you go. Each time you add a few streams or unit operations converge and save the model before moving on. Make sure you justify any additions that you make to the model. You can most likely model your system without including every unit from your PFD in the simulation. When modeling, think about the goal of your simulation and not about literally translating every real-world piece of equipment into the simulation model. Make the judgment call on what is most important and what the purpose is of every unit you add to your simulation. For example, if your real-life process shows a heat exchanger, a pump and a splitter, but all that is necessary is that the temperature, pressure, and separation are achieved properly, consider using a flash drum to accomplish all three.

8. Specifications within process simulation are everywhere and every stream and unit operation you place in your model contains specifications. In its simplest form, a specification is a constraint that you place on the model that the model needs to meet in order to successfully converge. For streams, these are usually temperatures, pressures, flowrates and compositions. For unit operations, they range from process conditions, like temperatures and pressures, to product conditions, like an overhead composition in a column. There are some pitfalls, however, that you can run into with specifications such as azeotropes in columns, high purity specifications, and the possibility of multiple solutions.

9. Petroleum assays are sets of distillation data that are used to characterize all of the components that reside in a given crude sample. When you enter distillation data into a process simulator, the program converts it into components that it can use when modeling your process. The components that are generated are called pseudo-components, and each pseudo-component represents all of the unknown components that would be within a given boiling point range in the crude. Some of the things you will want to ensure when properly developing your assay components are did you use the correct distillation curve basis, did you enter the data correctly, did you use enough cuts and are you using the correct characterization methods .

10. When you have more than 1 assay in a simulation model, you need to be concerned with assay blending and how it will affect the results of your file. Assay blending is the mixing of 2 or more groups of components (i.e. ¨C sets of cuts/pseudo-components) in order to calculate averaged component properties for cuts that are in the same boiling point range as one another. Once these new averaged component properties are determined, the program will use these averages from this point on in the simulation model any time it needs to access component

properties for a given pseudo-component. All component properties are blended, including normal boiling points, specific gravity, molecular weight and refinery inspection properties, and usually the blended component properties are calculated using a weighted average

These topics are based on experience and feedback from end-users to technical support representatives for process simulators. They often represent the most commonly asked questions or areas that most users find to be problematic. While the specific examples highlighted in each topic come from SimSci software, the intent of this tutorial presentation is to be broad-based and is particularly applicable to young professionals seeking to develop their experience with simulation.

Polymeric Sulfonic Acid Composites for Fuel Cell Applications: Effects of Casting Solvent and Sonication Time on Membrane Performance

Jimoh K. Adewole[1], *Abdullah S. Sultan*[1, *], *Amir Al-Ahmed*[2], *and S. M. Javid Zaidi*[3]

[1]*Center for Petroleum & Minerals, King Fahd University of Petroleum & Minerals, Dhahran 31261, Saudi Arabia*

[2]*Center of Research Excellence in Renewable Energy, King Fahd University of Petroleum & Minerals, Dhahran 31261, Saudi Arabia*

[3]*School of Chemical Engineering, The University of Queensland, Australia*

Abstract

Polymer blends comprising sulfonated poly (ether ether ketone) (SPEEK) and 3wt% (2-acrylamido-2-methyl-1-propanesulfonic acid) (PAMPS) were prepared by solution casting method. Two solvents (N,N-Dimethylacetamide (DMAc) and N-Methyl-2-pyrrolidone (NMP)) were uesd separately. Residual solvents in the membrane samples after drying were quantified using thermo gravimetric analysis. Effects of periods of sonication (0 to 90 minutes) and types of solvent on the membrane performance were evaluated interms of morphology and water uptake. An attempt was made to correlate membrane performance with the time of sonication in order to select the most suitable solvent and membrane casting conditions for subsequent membrane preparations. For both solvents, the water uptake was observed to first decrease at 10 minutes of sonication and then increased as the sonication time was increased. The minimum water uptake recorded for DMAc samples was 19.23% at 20 minutes sonication time while that of NMP sample was 24.24% at 10 minutes. However membrane morphology was found to deteriorate with increase in sonication time. No correlation was obtained between the membrane properties and the sonication time. In general, membrane performance was better for membranes samples using DMAc than NMP solvent. Therefore, DMAc was selected as the casting solvent for membrane preparation on which further studies will be carried out.

Keywords: morphology, water uptake, residual solvent, thermogravimetric analysis

Introduction

Water soluble polymers have received academic and industrial interests due to their potential applications in fuel cells and other electrochemical devices (Ye et al. 2012). One of the water soluble polymers that have attracted publication interest is poly (2-acrylamido-2-methyl-1-propanesulfonic acid) (PAMPS) (Florjańczyk et al. 2003). Generally, some of the major performance criteria for fuel cell materials include proton conductivity, water and methanol uptake, good mechanical properties, chemical stability, thermal stability and methanol permeability (Gil et al. 2004; Swier et al. 2005; Ye et al. 2012). PAMPS was reported to posses higher proton conductivity than partially hydrated Nafion. In addition, it has been found to be more tolerant to water fluctuation in high temperature operations. Some of the major drawbacks in the use of PAMPS are its poor mechanical properties, thermal stability, water and methanol uptake and solubility in many solvents (Florjańczyk et al. 2003). One of the approaches that have been proposed to solve these problems is development of polymer composites.

This research paper investigates the synergistic effect of combining water soluble PAMPS with sulfonated poly (ether ether ketone) (SPEEK) which is non - water soluble polymer to prepare composite membranes for application in fuel cells. Specifically, preliminary results on the performnance of the composite were presented. Performance were evaluated based on the composite morphology and water uptake. Effects of sonication time and types of casting solvent on membrane properties were also evaluated.

Material and Method

Material

Sulfonated poly (ether ether ketone) (SPEEK) purcahsed from FuMA-Tech, Germany, and poly (2-acrylamido-2-methyl-1-propanesulfonic acid) (PAMPS) purchased from Aldrich Chemical Co., are the two major polymers used in this research. The SPEEK has a DS of 0.533 and solubility of 99% in DMAc. SPEEK is derived by sulfonation of Poly (ether ether ketone) (PEEK). PEEK is a commercially successful polymer due to its excellent thermal stability and solvent resistance (Mohr et al. 1991). For sometimes, further application of PEEK have been stalled due this solubility issues. By sulfonation, PEEK can be made soluble in most commercial solvents. Importantly, the solubility of SPEEK also depends on its degree of sulfonation (DS). Since, this work involves the use of two polymers, a solvent which will disolve both the PAMPS and SPEEK must be selected. Literature review showed that PAMPS dissolves in N,N-Dimethylformamide (DMF), N,N-Dimethylacetamide (DMAc), propylene carbonate (PC), and water (Florjańczyk et al. 2003). Initial solubility studies that was carried out in our laboratory was done for N,N-Dimethylacetamide (DMAc) and N-Methyl-2-pyrrolidone (NMP). The choice of NMP was due its unique solvent properties and its structtural distiction from DMF and DMAc (Figure 1). Moreover, no literature was found where NMP was used as casting solvent for PAMPS. Thus, DMAc and NMP were used as solvents in this work. All materials were used as received from the manufacturers.

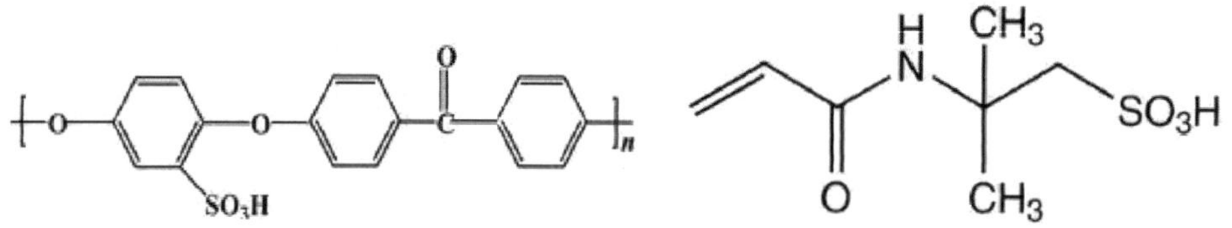

Figure 1 Structure of SPEEK and PAMPS

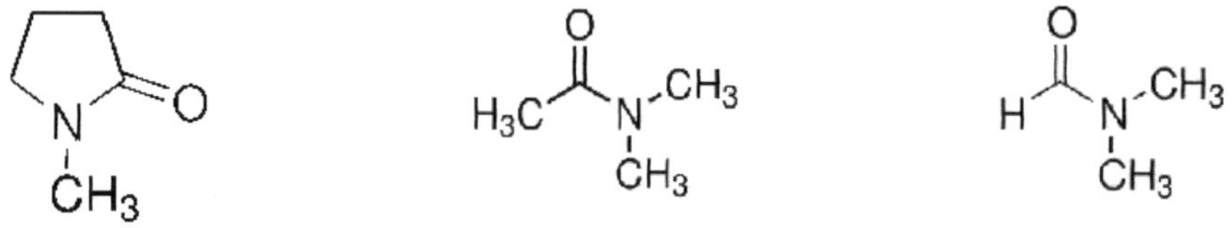

N-Methyl-2-pyrrolidone N,N-Dimethylacetamide N,N-Dimethylformamide

Figure 2 Structure of Solvents

Experimental Procedure

Membrane Preparation

SPEEK was first dissolved in solvents (DMAc and NMP) and stirred for 2hours at 70°C. PAMPS was then added and the blend was stirred for 24hours. The solution was sonicated for various periods of time ranging from 10 to 90minutes using an Ultrasonic bath. In untralsonic bath, the final temperature achieved for a given time of sonication depends on the tempearutre of the laboratory (summer or winter). It is therefoe genrally advisable that maximum temperature of the bath is determined and maintains (as the so - called equilibrium tempearture), when operating continuously under ambient conditions (Santos et al. 2009). During the sonication, equilibrium temperature was maitained by simply filling the bath with water that was preheated to sample temperature.

Membrane Characterization

Water uptake and morphological tests were performed. Water uptake was carried out by first vacuum drying all membrane samples at 100°C until constant weights were obtained. Dried samples were then immersed in deionized water for 24h at room temperature (25°C). Samples were taken out of the water after 24h, wipped to remove surface water and then weighed immediately. Water uptake was calculated using Eq. 1.

$$water-uptake=\frac{W_{wet}-W_{dry}}{W_{dry}} \qquad \text{Eq. 1}$$

Surface morphology of the membrane was done using scanning electron microscopy (SEM). Thermogravimetric analysis (TGA) was done from room temperature to 350°C using TA Q600 analyzer at heating rate of 20°C/min under nitrogen atmosphere with flow rate of 20mL/min using ~ 8 -10mg membranes samples.

Results and Discussion

Membrane Morphology by Scanning Electron Microspcopy (SEM)

Results of surface morphological analysis performed on membrane were shown in Figure 3 and 4. The overall pictures of this figure depicts a dense membrane with some surface defects observed for most of the sonicated samples. For example, Figure 3A showed some kind of layered arrangement with no visible pores at the surface. Also, the PAMPS seems to be well distributed throughout the SPEEK. The micrograph in Figure 3B can be observed to be rough and possesses some uneven ramdom distribution of PAMPS agglomerates indicating that the distribution of PAMPS have been disturbed by the 10minutes sonication. An increase in sonication time to 20 minutes was observed to improve the distribution of PAMPS. This was however accompanied by some cracks believed to be due to the slight increase in the viscousity of the blend solution. The vicousity increase was brought about by the excape of some fraction of the solvent during sonication. As the sonication time was increased, PAMPS was observed to become more evenly distributed and the cracks become more visible and larger in size (Figure 3D to F). SEM images of samples at sonication time of 10 and 30 minutes were reproduced in higher magnification of 10μm (Figure 4) in order to further revealed the effect of sonication on the morphology of the somposites. Similar morphological behaviour has been reported in literature, though for inorganic/organic clusters of nanoclay/epoxy composites membranes (Lam et al. 2005).

Water Uptake

Water uptake have been reported to be affected by the degree of sulfonation, membrane pretreatment, hydration, ambient relative humidity and water temperature (Xing et al. 2004; Wu et al. 2006). Figure 5 showed the effect of sonication time and solvent type on water uptake that was measured at room temperature (25°C). For both solvents, water uptake is same at zero sonication time. There was however a further decrease in the water uptake for NMP at 20 minutes sonication time before it beging to increase. Thus, the best sonication time for both solvent lied between 10 and 20 minutes with respect to water uptake. Interestingly, this result correspond to that reported by Lam (2005) on the effect of ultrasonication on microhardness of composites.

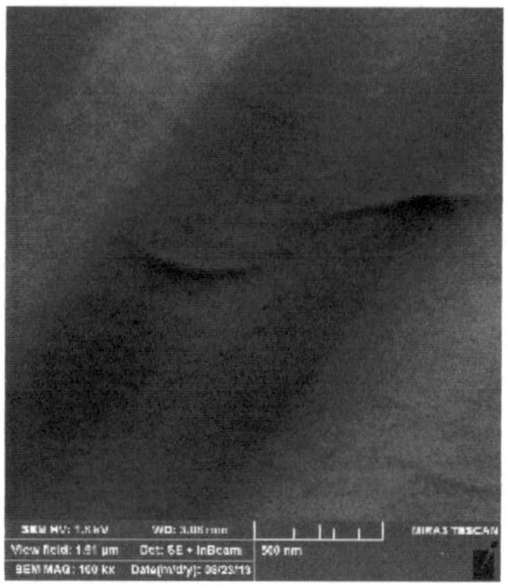

A) No sonication

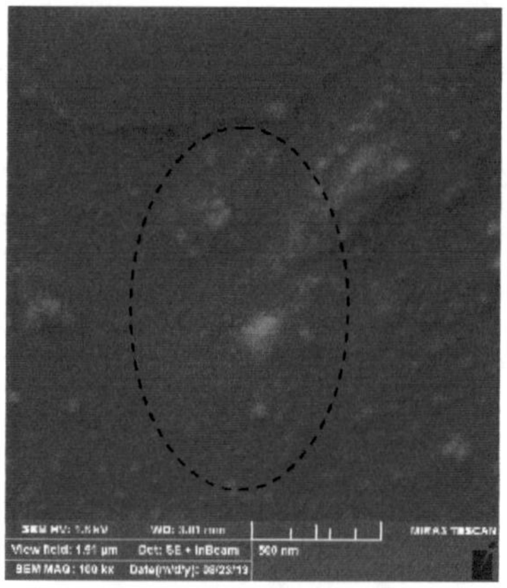

B) 10 minutes sonication

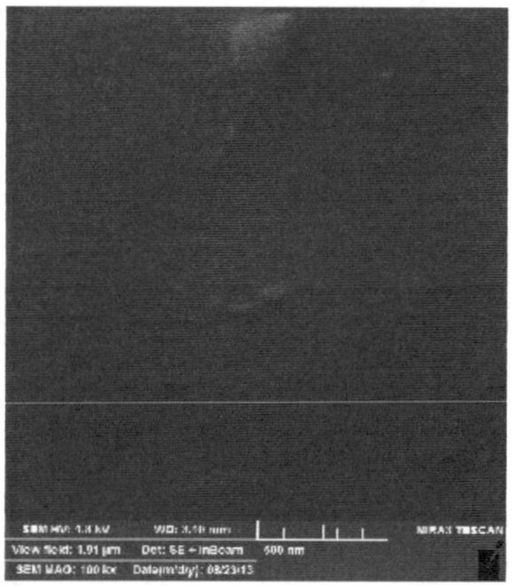

C) 20 minutes sonication

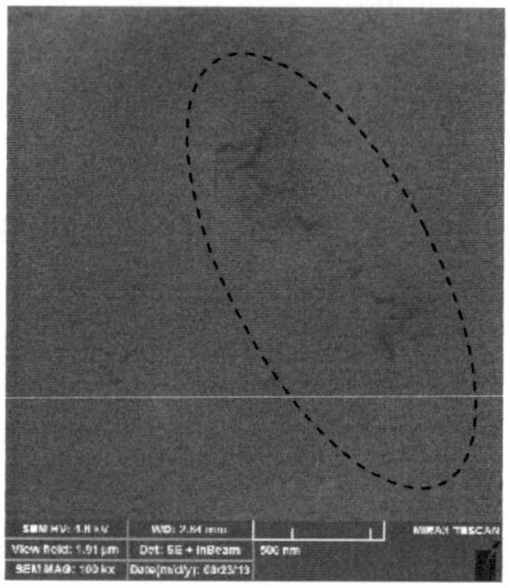

D) 30 minutes sonication

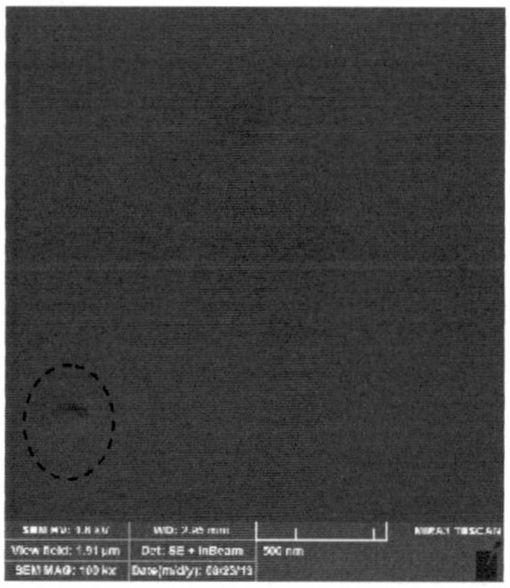

E) 60 minutes sonication

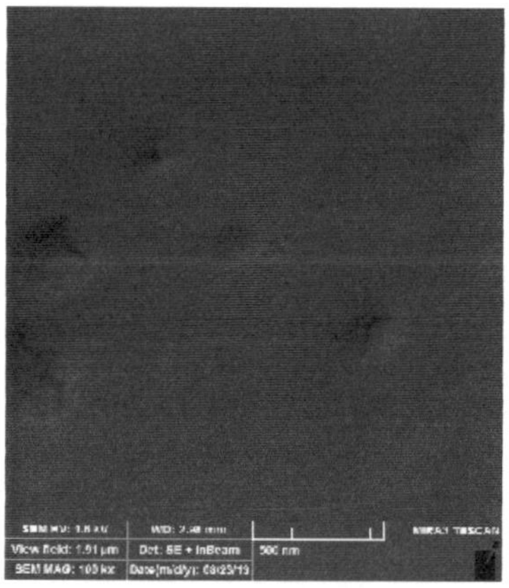

F) 90 minutes sonication

Figure 3 SEM micrograph of SPEEK-PAMS composite with various times of sonication with magnification of 500nm

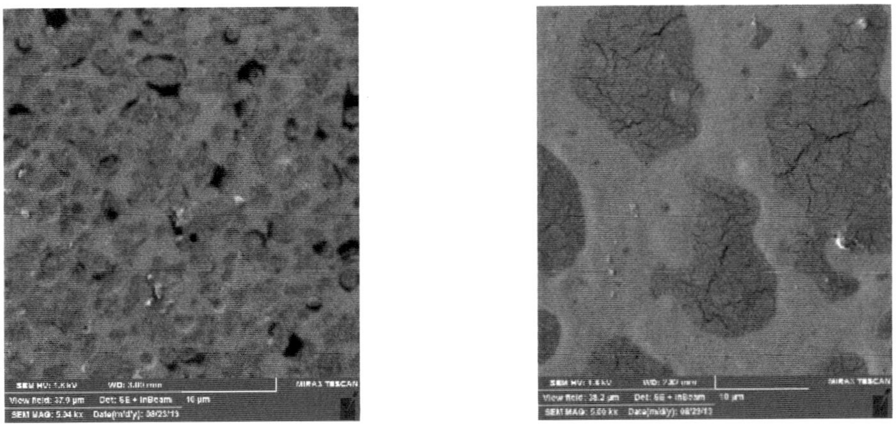

A) 10 minutes sonication B) 30 minutes sonication

Figure 4 SEM micrograph of SPEEK-PAMS composite with various times of sonication with magnification of 10μm

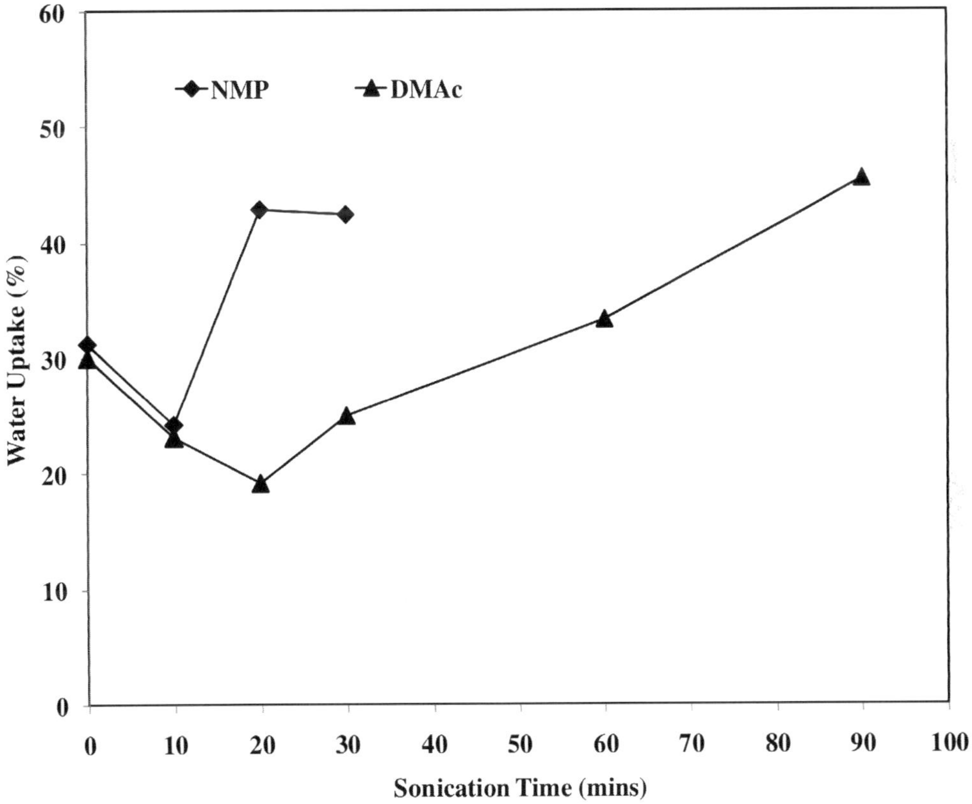

Figure 5 Water uptake of SPEEK-PAMS (3wt%) composite with variuos sonication times

Thermogravimetric analysis (TGA)

The difference in membrane performance resulting from the use different casting solvents have been attributed to the presence of residual solvents in the membrane samples after evaporation (Joly et al.

1999). TGA was thus used to quantify the residual solvent and evaluate its effect on the membrane performance as has been previously demonstrated in lieterature (Recio et al. 2007; Iqbal et al. 2008). Figure 6 shows the TGA curve for samples prepared from NMP and DMAc with different post vacuum heat treatment conditions. From this figure, weight loss can be attributed to evaporation of variety of substances, including solvents, from the samples (Rich et al. 1997). Table 2 contains the solvent and its boiling points, and weigth loss at the boiling point range that were extracted from TGA results. For sample heat treated for 24hrs, the overall percentage weight loss up to 204°C was about 8wt% of membrane sample for NMP while it is about 6.3wt% for DMAc. These will definitely include compnenents such as water, solvents and other unidentified substances. To quantify residual solvents, weight loss was estimated between 192 -214°C for SPEEK-PAMPS-NMP and between 154 and 176 for SPEEK-PAMPS-DMAc. The results obtained showed that the amount of the residual solvent in NMP is larger than that of DMAc by 1.25wt% (Table 1). To further investigate if this difference is responsible for higher value of water uptake for the NMP sample, membrane samples made from both solvents were further heat treated under vacuum for 72hrs. Residual solvent after this treatment was slightly higher in DMAc than NMP. Despite this, the value of water uptake of the NMP membrane is still higher than that of DMAc by approximately the same value as before the heat treatment (1.71wt%).

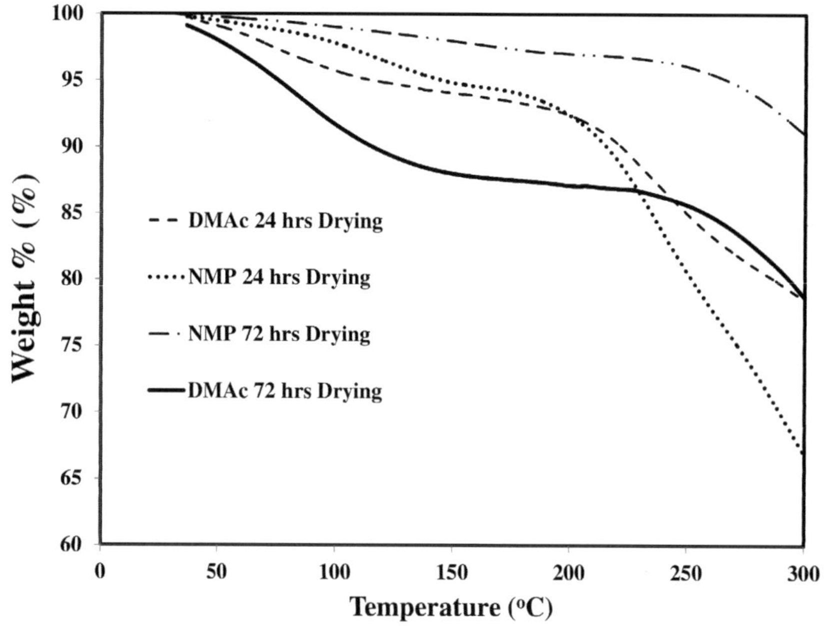

Figure 6 TGA of SPEEK-PAMPS-NMP and SPEEK-PAMPS-DMAc

Table 1 Residual solvent in SPEEK-PAMPS composites membrane samples

Membrane Sample	Solvent	B. Pt. (°C)*	Residual Solvent (wt%)	Water Uptake
DMAc 24 hrs Drying	DMAc	164.5 – 166	0.57	30.00
NMP 24 hrs Drying	NMP	202 – 204	2.85	31.25
DMAc 72 hrs Drying	DMAc	164.5 – 166	0.367	21.21
NMP 72 hrs Drying	NMP	202 – 204	0.217	22.92

*Boiling Points of Solvents

On a general note, results from Figire 1 – 5 showed that the best perfomance can be obtained for variety of sonication time depending the property of preference. Also, performance of DMAc solvent was observed to be better than that of NMP on the basis of water uptake. For this reason, DMAc was selected as the casting solvent for membrane preparation on which further studies will be carried out. Water uptake of DMAc membrane, prepared without sonication, was comapred with those reported in previuos literature. Table 2 illustrates the comparison between SPEEK-PAMPS and other relevant membranes previuosly reported in literature. In addition, this table serve to provide insight into the possible potentials of the composites. From this table, it was observed that the water uptake for SPEEK-PAMPS is within the range of values of other membranes on the table. Its water uptake is less than that of Nafion 112 which is a commercial membrane. The differences observed between the water uptake values of SPEEK – PAMPS and other membranes from the table can be explained based on three important factors - SO_3H group, the degree of sulfonation and experimental temperature (Gil et al. 2004; Shan et al. 2006; Wu et al. 2006). Water uptake of SPEEK – PAMPS is higher due to the presence of SO_3H group in the PAMPS molecules (Figure 2). SO_3H group have been reported to enhance membrane acidity and hydrophilic properties (Shan et al. 2006). Infact, the higher the SO_3H group the larger the water uptake. The degree of sulfonation has been reported to affect the water uptake of membranes. The higher the degree of sulfonation, the larger the value of water uptake (Gil et al. 2004). Thus, water uptake of SPEEK – PAMPS is lower than SPEEK -3 due their different degree of sulfonation, 53.3 and 72% respectively and the experimental temperature 24 and 60°C, respectively. The experimental temperature of SPEEK – 2 is close to that of the composite. However, the results cannot be compared in this case since the DS of SPEEK – 2 was not known. Above explanation is also true for the values of water uptake for SPEEK – 1.

Table 2 Comparison of water uptake values obtained with those from Literature (Xing et al. 2005)

Membrane	Water Uptakc (wt%)	Expcriment Temperature (°C)	**D S (%)	Reference
Nafion 112	22.85	N/A	-	(Xing et al. 2005)
SPEEK – 1*	16.00	30	45	(Ren et al. 2005)
SPEEK – 2*	44.90	25	N/A	(Zhong et al. 2008)
SPEEK – 3*	85.00	60	72	(Wu et al. 2006)
SPEEK– 4*	15.00	N/A	N/A	(Handayani and Dewi 2011)
SPEEK/PAMPS (3wt%)	21.21			This work

*The numbering have been included for explanation purposes.
**DS means Degree of Sulfonation

Conclusion and Recommendations

Blend of SPEEK and PAMPS was prepared by solvent casting method and its performance was evaluated with respect to its morphology and water uptake. Effect of sonication time (10 – 90minutes) and solvent types (DMAc and NMP) were also investigated. Performance of membrane prepared using DMAc was observed to be better than that of NMP. There was no correlation between the sonication time and membrane properties, but it was observed that the sonication deteriorate the membrane morphology. Membrane prepared without sonication shows better surface morphology, whereas the best water uptake was observed between 10 and 20 minutes. Thus, other properties need to be investigated in order decide the best sonication time. Therefore, further investigation such as thermal stability, methanol permeability, ion exchange capacity and gas (O_2 and H_2) permeability of the membrane are recommended for a complete evaluation of the membrane performance for direct methanol fuel cell (DMFC) application. All these investigations are in progress in our laboratory.

Acknowledgement

Authors would like to thank the National Science, Technology and Innovation Program (NSTIP) of KACST for funding this research under project number 10-ENE1374-04. Also, authors extend their appreciation to Center of Petroleum & Minearls, Center of Excellence in Renewable Energy, Mr. Dan Jerry D.Cortes of the Nanotechnology Laboratory, Chemical Engineering Department, King Fahd University of Petroeleum & Minerals, Saudi Arabis for their supports.

REFERENCE

Florjańczyk, Z., et al. (2003). "Proton conducting electrolytes based on poly(2-acrylamido-2-methyl-1-propanesulfonic acid)." Electrochimica Acta **48**(14–16): 2201-2206.

Gil, M., et al. (2004). "Direct synthesis of sulfonated aromatic poly(ether ether ketone) proton exchange membranes for fuel cell applications." Journal of Membrane Science **234**(1–2): 75-81.

Handayani, S. and E. L. Dewi (2011). "Influence of Silica/Sulfonated Polyether-Ether Ketone as Polymer Electrolyte Membrane for Hydrogen Fueled Proton Exchange Membrane Fuel Cells." Internat. J. of Sci. and Eng. **2**(2): 26-30.

Iqbal, M., et al. (2008). "Solvent effect on morphology and CO2/CH4 separation performance of asymmetric polycarbonate membranes." J. Membr. Sci. **318**(1–2): 167-175.

Joly, C., et al. (1999). "Residual solvent effect on the permeation properties of fluorinated polyimide films." Separation and Purification Technology **16**(1): 47-54.

Lam, C.-k., et al. (2005). "Effect of ultrasound sonication in nanoclay clusters of nanoclay/epoxy composites." Materials Letters **59**(11): 1369-1372.

Mohr, J. M., et al. (1991). "Gas transport properties of a series of poly(ether ketone) polymers." Polymer **32, 13**: 2387-2394

Recio, R., et al. (2007). "Gas separation of 6FDA–6FpDA membranes: Effect of the solvent on polymer surfaces and permselectivity." J. Membr. Sci. **293**(1–2): 22-28.

Ren, S., et al. (2005). "Surface modification of sulfonated poly(ether ether ketone) membranes using Nafion solution for direct methanol fuel cells." Journal of Membrane Science **247**(1–2): 59-63.

Rich, D. C., et al. (1997). "Effect of cure conditions on Probimide 32 polyamide-imide." Journal of Applied Polymer Science **63**(9): 1113-1126.

Santos, H. M., et al. (2009). The Power of Ultrasound. Ultrasound in Chemistry: Analytical Applications. J.-L. Capelo-Martinez. Weinheim, Wiley-VCH.

Shan, J., et al. (2006). "Sulfonated polyether ether ketone (PEEK-WC)/phosphotungstic acid composite: Preparation and characterization of the fuel cell membranes." Pure Appl. Chem **78**(9): 1781 - 1791.

Swier, S., et al. (2005). "Sulfonated poly(ether ketone ketone) ionomers as proton exchange membranes." Polymer Engineering & Science **45**(8): 1081-1091.

Wu, H.-L., et al. (2006). "Preparation and characterization of poly(ether sulfone)/sulfonated poly(ether ether ketone) blend membranes." European Polymer Journal **42**(7): 1688-1695.

Xing, D. M., et al. (2005). "Characterization of Sulfonated Poly(Ether Ether Ketone)/Polytetrafluoroethylene Composite Membranes for Fuel Cell Applications." Fuel Cells **5**(3): 406-411.

Xing, P., et al. (2004). "Synthesis and characterization of sulfonated poly(ether ether ketone) for proton exchange membranes." Journal of Membrane Science **229**(1–2): 95-106.

Ye, Y.-S., et al. (2012). "Water Soluble Polymers as Proton Exchange Membranes for Fuel Cells." Polymers **4**(2): 913-963.

Zhong, S., et al. (2008). "Modification of sulfonated poly(ether ether ketone) proton exchange membrane for reducing methanol crossover." Journal of Power Sources **180**(1): 23-28.

Fixed Plant Facility Siting – A Mega Project Lifecycle Framework

M.F. Haider
mech_haider@yahoo.com
mhaider@suncor.com

Prepared for Presentation at
American Institute of Chemical Engineers
2014 Spring Meeting
10th Global Congress on Process Safety
New Orleans, LA
March 30 – April 2, 2014

Fixed Plant Facility Siting – A Mega Project Lifecycle Framework

M.F. Haider
mhaider@suncor.com
mech_haider@yahoo.com

Keywords: Facility Siting, Project Management, Mega Project, Consequence Analysis (CA), Quantitative Risk Assessment (QRA), Adaptive, Change driven, Agile methodology

Abstract

Facility Siting based on tools such as Consequence Analysis (CA) and Quantitative Risk Assessment (QRA) is often times successfully applied for Brownfield operating unit projects, projects with relatively smaller footprints and for projects lasting relatively short time-periods. Challenge remains to apply and realize the full potential and value of Facility Siting for Mega-Projects. This is mostly due to project complexity and length – this paper will introduce a Management Framework to successfully apply CA and QRA as part of Facility Siting for Mega-Projects. The paper will also outline key strategies to ensure successful (i.e. value adding) implementation of facility siting, suggest a list of activities, associated sequence and pitfalls to avoid. Project Management team (e.g. Project Sponsor, Project Engineers, Managers) of both owner/operator and contractor/vendor may find this paper more useful as it provides a broad overview (i.e. when to initiate, how to plan, execute and close such assignment to maximize value) including real life examples of facility siting process over the project lifecycle.

1. Introduction

The driving force for process safety has been primarily based on catastrophic events [1]. With an increasing number of incidents, the process industry and governments started to take measures such as Facility Siting initiatives with hope to minimize loss of life and property. In the 1970s and 1980s, some of the world's most tragic industrial accidents took place. The Flixborough explosion in 1974 was reinforced by that of the Seveso tragedy in 1976 [2]. The Macondo blowout in the Gulf of Mexico on April 20, 2010, killed 11 employees and led to an uncontrolled oil spill lasting 87 days. This blowout was the most significant offshore incident in the US to-date. Finally, the Fukushima Daiichi nuclear plant incident in Japan in March 2011 drew the attention of the global process and power industries and encouraged them to incorporate natural disaster risks in a hazard analysis study [3]. As part of inherently safer design, many companies now-a-days apply CA and QRA as part of facility siting for their new projects. Due to the scarcity of information at the beginning, the studies in many cases start later in the project lifecycle for them to be effective and value adding. This paper will focus more on how to manage such initiatives for new Mega-Projects over the technical intricacies of the CA/QRA processes. The intention of the paper is to enhance the Body of Knowledge publicly available for management team in as much non technical languaga as possible to make their greenfiled Mega-Project initatives a success i.e. safer for people, environment and property.

So, what is different for industrial Mega-Projects?

- Industrial Megaprojects, as the name implies, are very large. To provide a simple and simply applied definition, we are defining a megaproject as any project with a total capital cost of more than $1 billion (U.S. dollars) as measured on January 1, 2003 [4].
- Most often engineering and construction activities run in parallel and not in sequence
- Multiple sub-projects under the 'Mega-Project umbrella' run in parallel
- Complex – design is almost 'never' complete in early stages – goes through many iterations
- Progressive elaboration is evident
- Scarcity of critical workforce with mega-project experience
- Scarcity of information – requires project team to make assumptions for CA/QRA
- Timeline: 2-5 years

2. Key Strategies

2.1 Facility Siting Charter

It is not only important or good to do but critical that the stakeholders need to identify, understand and agree on the purpose and value to conduct facility siting for their projects. The Author's personal experience over the years has been in order to maximize value of Facility Siting the project team has to start 'early' to develop project Charter and start with the end in mind. As can be seen from the following generalized Figure 1 that as the company starts to appraise opportunities, it is

Figure 1: Stages and Milestones to apply CA/QRA in a Mega-Project

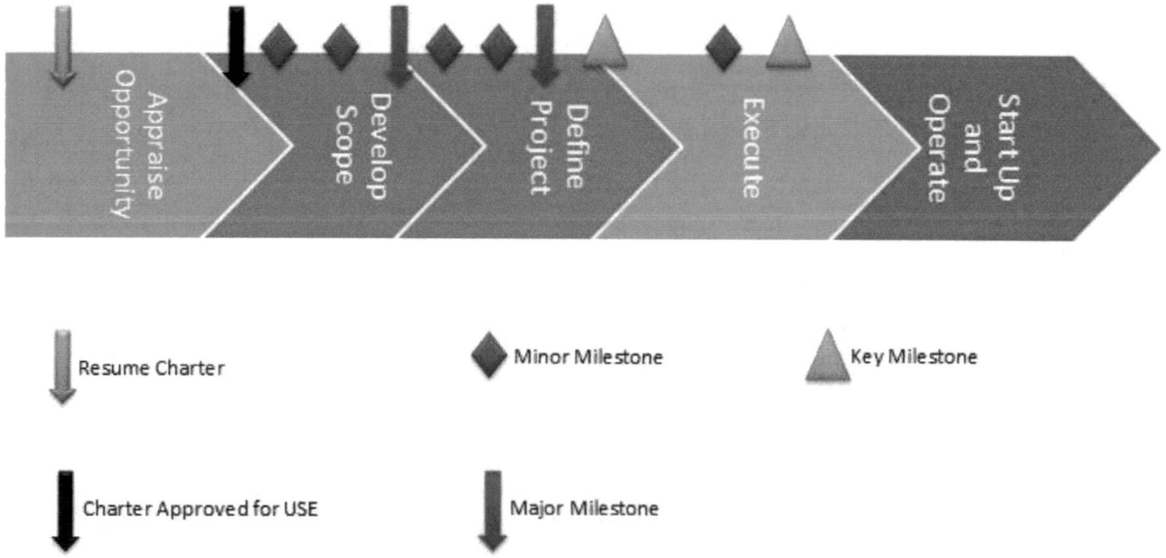

prudent to resume developing the Facility Siting Charter by bringing in the stakeholders together. Figure 1 shows a generalized project lifecycle for a typical Petro-Chemical Mega-Project. The

Figure also shows that the Key Deliverable is developed through a successive series of repeated cycles over the project life-cycle. The Milestones will be defined later in the paper. Part of starting with the end in mind is to clearly outlining ultimate goal/s, e.g. vision is to design a safer facility and utilize Mitigated QRA Report to formally document and communicate that facility is within the Risk Threshold or justification for accepting risk profile as is by Senior Management. The Vision in the Charter also provides opportunity to strike a balance between business goals and climbing the Risk Mountain – at each stage of increasing safety (decreasing risk). As part of striking a balance, the Project Engineer has to Identify options up-front if and/or when it becomes apparent that aggregate Residual Risk will not be within the Risk Threshold or acceptable so that team can put more effort in developing justification for either accepting or rejecting the risk profile by Management

Figure 2: Strike a balance between Business goal and Climbing the *expensive* Risk Mountain

2.2 Management Strategy

In order to successfully manage Facility Siting, CA and QRA for a Mega-Project, the stakeholders have to align on a management strategy. There are two primary management approaches for the purpose of managing such work:

1. Predictive or Plan driven
2. Adaptive or Change driven

The Author's past experience has been if the organization undertook very 'similar' projects in the near past (<5 years), critical experienced workforce is retained, Lessons Learned is readily available, and most importantly, the past project is still considered to be a success, it is advisable to use Plan driven approach. On the other hand, if the organization is introducing new technology, has limited past experience with similar project, involves relatively high number of stakeholders, capital cost > $1B, project life spans >2 years, it is recommended to utilize Change driven approach to maximize value. It is to be noted that it is not essential to meet all the suggested criteria mentioned above to pick one management strategy over the other.

2.2.1 Benefits of Adaptive or Change Driven Approach

In Adaptive/Change Driven Planning, project team intentionally repeats one or more project activities as the project team's understanding of the product increases ……….Also known as …agile methods …..intended to respond to high levels of change and ongoing stakeholder involvement [5]. The heart of Facility Siting involves making timely but informed decision. Some of the characteristics of this framework:

1. To add small incremental improvements that will deliver value to stakeholders
2. Milestones develop the product (e.g. Mitigated QRA Report) through a series of repeated cycles, while increments successively add to the functionality of the product
3. For each Milestone, a deliverable (read 'model') or set of deliverables will be completed. Future iterations may enhance those deliverables or create new ones. Each iteration incrementally builds the deliverables until the exit criteria for the phase are met, allowing the project team to incorporate feedback.

The biggest challenge with Mega-Project is scarcity of key information needed as inputs to develop scenarios and models. Some of the critical benefits, a project may gain by applying Change Driven approach:

1. The approach acts as Leading Indicator on Process Safety for Plant Layout - allows for multiple opportunities to make critical decisions on plant layout over the project life-cycle capitalizing on what is known at a given time. This lets the project team to use "what they know", to ensure "what they don't know" falls into the right place
2. Opportunity to make decisions over the project lifecycle not only works to save life but also helps with striking a balance between business needs and climbing the risk mountain
3. Aides design-development with progressive elaboration while producing final CA/QRA deliverable for the project

2.3 Execution Strategy

The main idea of Change Driven process starts with decomposing the known preliminary layout and process information into a set of requirements, work to be performed and their priority sequence, sometimes referred to as a *Product Backlog*.

It is advisable that the owner/operator forms a multi-use agreement (contract) with the chosen vendor. This way the plant areas under the Mega-Project umbrella can get their specific work done while the CA/QRA Project Team provides overall oversight and guidance.

At the beginning of an iteration, the team will work to determine how many of the highest priority items on the backlog list can be delivered within the next iteration. The time-length of each iteration should be less than 4 weeks. As shown in Figure 1, during "Develop Scope" and "Define Project", the length of each iteration could be even less than 2 weeks. This iterations (otherwise known as "Minor Milestone" as shown on Figure 1) will develop scenarios, assumptions and preliminary CA/QRA models. Assumptions are required due to lack of available quality information in the early stages of a project.

At the end of each iteration, the Minor Milestone/deliverable should be ready for review by the customers/stakeholders. This does not mean that the customer is required to accept delivery, just that the product should not include unfinished and/or incomplete features.

The sponsor and customer representatives should be continuously engaged with the other project stakeholders to provide feedback to Contractor on deliverables as they are created and to ensure that the *Product Backlog* reflects Sponsor/stakeholders' current needs. Below, you will find two examples using two different management strategies. For confidentiality purpose, the specific name/location is kept generic and few items were modified to suit.

2.3.1 Project X Facility Siting Using Change Driven Approach

A CA/QRA team was given the assignment to support design development during "Develop Scope" phase (Figure 1). The Sponsors already selected the site for the project. But the fixed plants were yet to be located on the site. The Vision for the initiative was for the QRA result to come under the organization Risk Threshold. Multiple EPC's were given to engineer different plant areas. The CA/QRA team selected a vendor to support them with defining the *Product Backlog* based on the preliminary Process Design Basis, intended product the facility is expected to sell, the feedstock necessary etc. The CA/QRA team had prioritized the Product Backlog based on hazard level of feedstock, preliminary PHA schedule and project schedule. This also involved obtaining feedback from different Process Engineering and Operations teams to collect necessary inputs to put together the assumptions and scenarios to be modeled within the 1st iteration.

Out of 25+ scenarios, only 7 perceived to be the most hazardous were modeled for CA that gave ample input (e.g. explosion, toxic and fire contours) for the CA/QRA team to work with the stakeholders to recommend location for specific process units on the given site. Note that, CA/QRA team integrated the potential consequence to understand local incident and global consequence events. At this time QRA was not performed. This was used as input for minor modification of preliminary plant layout. It is to be noted that the site plan was still preliminary after the first iteration but assisted different EPC accountable for their specific area to recognize where to locate the specific plants i.e. finalized key plans for the overall site plan.

With further design-development, more refined information became available from each EPC. This new information led to 130+ scenarios and was used for further CA/QRA iterations that created either new model and/or embellished the past ones. Each EPC was accountable for facility siting within their boundary while the CA/QRA team provided technical oversight and integration support for the overall project. The Sponsor/stakeholders were engaged on a regular basis to ensure that the Product Backlog reflects future modeling requirements to properly site facilities e.g. Fire Water Pump house, Tank Farms, site Medical Centre etc. Early engagement of CA/QRA team and higher management willingness to incorporate their feedback with the progressive elaboration of the project led to QRA result meeting the organization risk threshold and ultimately safer facility. Changes suggested based on CA/QRA work in earlier stage (i.e. Develop Scope, Define Project) were easier to incorporate into the design.

2.3.2 Project Y Facility Siting Using Plan Driven Approach

The sponsors for project Y selected the site for the project and decided to hold on to not perform CA/QRA work with the progressive elaboration of the project activities. This was mostly decided based on the fact that not all information was available to properly conduct the 'studies'. Fundamentally, the sponsors looked CA/QRA as deliverables requiring a more 'complete' plan. Once the Develop Scope (as referred in Figure 1) is almost complete, the sponsors decided to engage CA/QRA team and a vendor. By this time, the project had significantly high number of stakeholders. Many of the stakeholders were not very familiar with CA/QRA and the value it could bring to them. There was not much time to schedule and train the stakeholders as well to bring them up to speed. As a result, the stakeholders were unable to align on a Vision. The Vendor was asked to provide explosion, toxic and fire contour maps for specific scenarios and locations that were almost 'fixed' on the site plan at that point in time. The resultant contour maps showed significant consequence on nearby facility with over 700+ people. It was also noted that the plant area consuming the most hazardous substance would impact the nearby facilities significantly from consequence perspective. It should be noted that the result was only based on a selected few scenarios. By the time the project reached the "Execution" phase, more scenarios were added as part of the scope. But it was already too late to incorporate most of the recommended changes resulting from CA/QRA work. The project went through several area by area risk reduction workshops only to accept (by the company executives per organization risk matrix) the aggregate residual risk level which was also found to be not within the organization risk threshold.

2.3.3 Importance of Starting Early

Figure 3: Impact of variable based on project time [5]

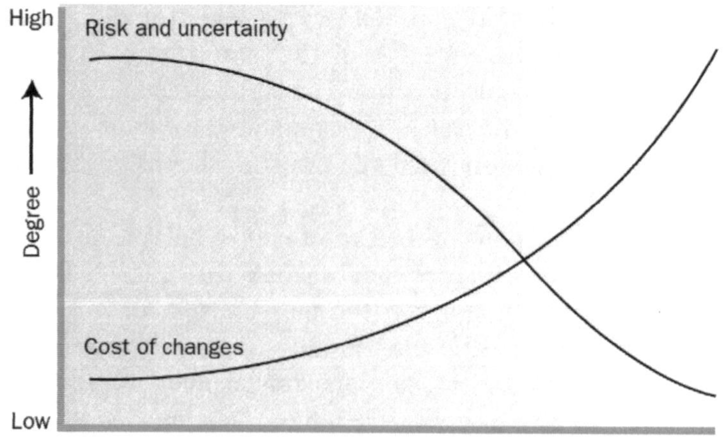

The above two examples highlight the importance of starting CA/QRA work early enough to realize and maximize the value CA/QRA brings to a Mega-Project. The high-level view provides common frame of reference for comparing projects for CA/QRA initiative, even if the projects were dissimilar. Change driven methodology allows for applying the known information at early stage in a way that supports laying out plant facilities to reduce risk and thus maximize potential value. While these characteristics remain present to some extent in almost all project life cycles, they are not always present to the same degree [5]. Change Driven framework, in particular, keeps stakeholder influences higher and the costs of changes lower especially in the early stages of the life cycle than in predictive/plan life cycles. The Figure 3 highlights a well-known fact of project management that even though risk and uncertainty remains high in the early stages of a project, cost of incorporating changes remains much lower. This translates into a key opportunity for project stakeholders to align on laying out the key areas on a site plan intelligently to reduce risk and maximize value by applying tools such as CA and QRA.

2.4 Key Decision Points

Facility Siting, from a management point of view, is all about bringing the stakeholders closer and find the 'right' spot for the facilities. Finding the 'right' spots requires making timely decisions. The challenge has always been to make timely decisions as the stakeholders more often find themselves forced to make those decisions based on information that are frequently of poor quality and at times uncertain, especially in the early stages of a project. At times, due to the scarcity of information, stakeholders rely on their past experience to make those decisions on plant layout. As a result, depending on the individual experience, two different teams for a project area may come up with two very different plant layout. CA/QRA work, if started early enough, will also depend on input that most often requires embellishment through assumptions to create needed scenarios and models. Even with this limitation, CA/QRA provides a key 'input' in allocating real estate for key facilities on a plant site at early stage of a project lifecycle. With each iteration and deliverable 'Change driven' methodology allows for multiple opportunities to make critical decisions on plant layout over the project life-cycle. It allows the stakeholders to decide when to 'sharpen the pencil' (e.g. PHAST to Computational Fluid Dynamics model), relocate facility or accept the consequence and/or risk, 'pull the plug' on further modeling/refinement and start developing justification, if required, to recommend management to accept/reject the aggregate risk profile.

2.5 Alignment with Process Hazard Analysis (PHA) Planning

CA/QRA results are a useful input to PHA (e.g. HAZOP, LOPA) activities. This is particularly true as projects move forward after establishing the plant layout on a high level and starts to conduct "Define Project" (Figure 1). PHA activities may also result in recommendations (i.e. new scenarios) that goes into the *Product Backlog* and requires project team to develop new models or refine old ones. Mega-projects have multiple sub-projects running in parallel most of the time, having an Integrated Schedule for all the reviews is a good strategy so that the right stakeholders are booked for the sessions. The integrated PHA schedule also supports prioritization of *Product Backlog* for CA/QRA activities over the lifecycle of the project.

2.6 Integration Strategy

It is critical for a Mega-Project involving multiple plant areas/silos to understand the BIG picture so that site risk can be dealt with on an aggregate basis. There is strong possibility of one area to impact other plants in the vicinity. It is recommended to review integration aspect and aggregated results on a regular basis. The following Figure 4 shows that the origin of an incident may reside in one particular plant area, while it may impact other areas in the vicinity.

Figure 4: Consequence of an incident in one area impacting multiple areas

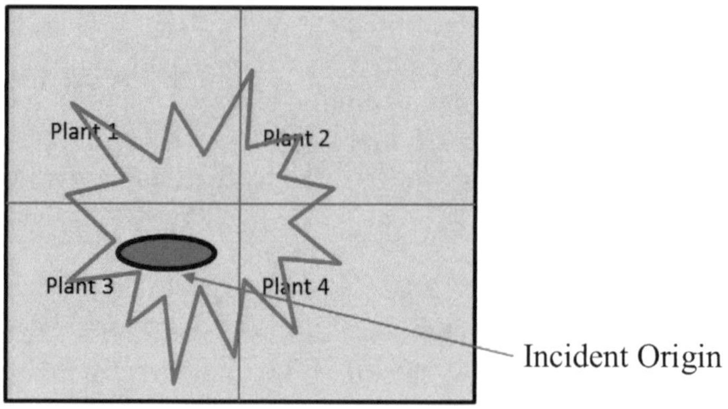

Recommended Strategy:

- Setup Strong Center - Task Force Single Point of Contact (SPOC)
- Collaborate – SPOC will promote Collaboration among Plants/Areas

3. List of Activities (suggested)

3.1 Overall Sequence

The CA/QRA work typically should start with the approved Charter as shown on Figure 1. A project Task Force i.e. SPOC is formed comprising of Project Engineer, Operations rep, Process rep and CA/QRA Technical Expert. The SPOC is ultimately accountable to produce the CA/QRA deliverable studies but also acts to support the design development and site plan finalization efforts in the early stages of the project.

The team collects 'input' such as Preliminary site Layout and design information, works with area process and operations team to identify scenarios, develop assumptions for known-unknowns and unknown-unknowns, develop modeling 'Product backlog', prioritize backlog based on PHA schedule and produce modeling output (i.e. deliverable) for stakeholder use. The stakeholder usually has the option of accepting or rejecting consequence. If the result of an iteration is acceptable within area causing the incident – communicate to outside stakeholders to ensure there is no cross boundary impact. If the result of the iteration is NOT acceptable to outside of the area – use the result as input to PHA for the areas to mitigate. Collaboration recommended between origin area and impacted area/s to identify appropriate mitigations. There is also a possibility that the consequence resulting from the iteration is not acceptable to the area causing the incident, in that case the area will initiate a PHA to mitigate the consequence. As part of developing mitigations, consider the following:

- Develop QRA if appropriate
- Perform Building Assessments for Critical Buildings
- Apply Cost-Benefit analysis to optimize risks/opportunities

- Determine Safe Haven locations and their characteristics

If Mitigation is not possible from Business perspective – perform As Low As Reasonably Practicable (ALARP) justification for Senior Management to formally accept the posed risk

- To apply mitigations on a site wide basis, review Milestone deliverables by taking on a BIG picture perspective
- Future iterations may enhance those deliverables or create new ones as new information is available. Each iteration incrementally builds the deliverables to produce the Mitigated CA/QRA report to support Facility Siting for the stage

The overall effort is expected to produce a maximum of two single reports for the overall site:

1. Mitigated version of QRA Report
2. and Consolidated CA Report

It is recommended to develop as-built, pre-start up CA/QRA reports prior to start-up. Table 1 and 2 outlines typical activities performed by SPOC and plant areas respectively once key process areas are established/finalized on Site Plan. Please note that the activities are project dependent and may warrant modifications to suit specific situation.

Table 1. SPOC Sequence of Activities to conduct Facility Siting

Initiative	Key Steps
Setup Facility Siting Team	Sponsor to support to setup a team (i.e. Project Engineer, Technical Expert, Operation, Process representative) to act as SPOC
Integration	Provide functional (technical and management) guidance to Areas
	Assess cross boundary impact/opportunity
	Communicate incremental outputs after each Area PHA/Risk Reduction
IFA versions (prior to "Define Project" stage)	Consolidated CA Report
	Mitigated and Integrated QRA Report
IFD CA/QRA (prior to "Execution" stage)	Re-validate IFA CA/QRA post PHA and vendor PHA
As-built versions	Post-oil version

3.2 Typical Area Activities

Table 2. Area (Typical) Sequence of Activities to conduct Facility Siting

Initiative	Key Steps
Plant 1 (Typical)	Perform Modeling and Calculations
	Conduct Risk Reduction Workshop (if appropriate) – Apply mitigations
	Incremental output for use based on reduced risk level
	Assess Buildings/facilities
	Identify Safe Haven/shelter Locations (if required)
	Estimate Area Project Scope impact
	Perform area scope updates/changes e.g. passive fire protection requirement
	Carry out QRA (if appropriate)
	Update Product Backlog
	Develop/update contract to support area specific activities

4. Pitfalls to Avoid

The following observations were noted by the Author during his time to manage Facility Siting scope:

1. Facility Siting does not replace good and common sense engineering practice e.g. locating a plant that uses high volume of hazardous material will significantly increase risks to other plants/areas (e.g. camp) within the vicinity
2. Evaluate effectiveness of 'Cost Reduction' philosophies as opposed to added risk e.g. different modularization techniques may decrease the plant footprint by reducing pipe lengths but makes the plant more compact and likely to increase severity of explosion consequence
3. Avoid developing models with hope to realize risk reduction
4. Sharpening the pencil for models most likely may not reduce risk
5. Avoid jumping into CFD or other similar methods without evaluating cost-benefit for undertaking such effort
6. Utilization of QRA results is much more controversial than the methodology. Watson [6] has suggested that QRA should be considered as an argument, rather than a declaration of truth. In this view – focus on accuracy over precision
7. CA/QRA Task Force should be part of Management of Change (MOC) process
8. Evaluate carefully before locating facilities e.g. tank farms close to main entrance/egress as part of Facility Siting

9. Keep the focus on VISION for the overall effort to avoid scope creep
10. Note the difference between Facility Siting as a Deliverable and tool for design-development activities

5. Conclusion

This paper presented Adaptive/Change driven framework to manage initiatives like facility siting/CA/QRA as part of Mega-Projects. Change driven framework does not mean "lack of planning or no planning". As demonstrated in the paper, this methodology starts with a vision and then develops the ultimate product both iteratively and incrementally when it is possible to define small incremental improvements that will deliver value to stakeholders. This allows making key decisions on laying out the facilities in early stages of a project rather than waiting for design to develop to a significant extent. Often times, project team wait too long for design to mature enough before applying CA/QRA to validate consequence impact and/or risk level. Most often, especially for Mega Projects, this leads to delays in executing the CA/QRA work properly at the right time to influence site layout. This delay in execution leads to scenarios when scope changes cannot be incorporated anymore. This translates into value erosion as project was unable to reap the full benefit of Facility Siting. The management framework recommended in this paper i.e. Change driven/adaptive framework allows for the stakeholders to start early in the project lifecycle to apply CA/QRA in laying out the plant facilities. The key is to start EARLY but START with the END in mind. Alignment among stakeholder on a VISION up-front is also important as this is expected to save life while striking a balance between business needs and climbing the Risk Mountain. Vision also allows avoiding scope creep while applying Adaptive/Change driven approach. It is highly recommended that stakeholders focus more on accuracy over precision. In the early stages of project lifecycle, quality information is scarce. Change driven/Adaptive framework gives you as the project stakeholder the opportunity to use "what you know" intelligently to ensure "what you don't know" falls into the right place.

6. References

[1] M. Mannan, A. Chowdhury and O. Reyes-Valdes, "A portrait of process safety: From its start to present day," College Station, 2012.

[2] M. Mannan, Lees' Loss Prevention in the Process Industries, Volumes 1–3 (3rd Edition), , 2005, Elsevier, 2005.

[3] K. McAndrews, "Consequences of Macondo: A Summary of Recently Proposed and Enacted Changes to US Offshore Drilling Safety and Environmental Regulation," in *Society of Petroleum Engineers, Americas E&P Health, Safety, Security and Environmental Conference*, 2012.

[4] E. W. Merrow, Industrial Megaprojects Concepts, Strategies and Practices for Success, New Jersey.: John Wiley & Sons, Inc., 2011.

[5] A Guide to the Project Management Body of Knowledge (PMBOK® Guide) – Fifth Edition, Newtown Square: Project Management Institute, Inc., 2013.

[6] S. Watson, "The Meaning of Probability in Probabilistic Risk Assessment," *Reliability Engineering and System Safety,* vol. 45, pp. 261-269, 1994.

348480 Global Energy and Transportation and Mobil Oil

Abstract:

Biofuels, Computational Methods, Engineering Economics, Polymers, Thermodynamics. Polymers-plastic, Thermodynamics- oil, Engineering Economics – the money it takes to finish the method of Global Engineering Research, Computational Methods – numerical input of analytical data, Biofuels – the transport fluids that utilized in more than one method from industrial usage to everyday living which enters the atmosphere. Exxon has completed a $600 million dollar deal for production of algae biofuel, amongst Mobil Oil has performed major research Global Energy and the effects of oil. Reading a most recent article on Mobil website.

348748 Characterization of Iron Phthalocyanine As the Active Material for Lithium Batteries

Abstract:

Results on the determination of the lithium-ion diffusion coefficient in the iron phthalocyanine (FePc) active material as a function of intercalation level, using a modified electrochemical impedance spectroscopy technique, will be presented. Also, developed thermodynamic functions such as the Gibbs free energy, entropy, and enthalpy changes for the lithium insertion reaction with FePc, derived from cell open-circuit voltage measurements as a function of intercalation level, are presented. Also presented are estimates of the maximum amount of lithium that can be intercalated into the FePc based on constant volume and temperature conditions.

349086 Design of a Free-Fall Reactor for Fast Pyrolysis of Waste Plastics

Abstract:

Thermal pyrolysis has earned significant reputation in the recovery of valuable chemicals, especially petrochemical based fuels from various materials, including biomass and plastics. Of many available technologies, *fast* pyrolysis was preferred in his study owing to its minimal heat and mass transfer limitations and better control on process conditions. The focus of this study is towards efficient conversion of waste plastics into gaseous products consisting primarily of alkanes, alkenes and aromatics. Lab scale experiments were performed using a micro reactor to determine favorable conditions, including reaction temperature and particle size that would result in maximum yield of ethylene. Using these results, a simple industrial size radiatively heated free-fall reactor was designed for the purpose of pyrolyzing waste plastics. The goal is to design a *fast* pyrolysis reactor that improves upon inefficiencies in traditional reactors while resulting in high yield of ethylene.

349732 Biosorptive Dehydration of Ethanol/Water Azeotropes Using Compound Starch-Based Adsorbent

Abstract:

Abstract: Recently, the interest in the production of bioethanol has greatly increased as a result of the increasing fossil fuel prices and associated environmental issues. Bioethanol is made from either grains or biomass and the broth contains 6 to 12 percent ethanol. In addition, water removal represents a serious problem as ethanol form azeotrope with water at atmospheric pressure. Distillation to water-free ethanol consumes 50 to 80 percent of energy used in the fermentation ethanol manufacturing process. Various techniques have been developed to break the azeotropic point, such as salting-out method, azeotropic distillation, extractive distillation, reactive distillation, adsorptive distillation and pervaporation. Among the methods, the adsorptive option is the most attractive, especially the biosorption, which has distinct advantages over conventional methods as it is environmentally friendly, highly selective, easily available, easy to operate, cost effective and reusable in repeated cycles in the treatment of aqueous organics.

In this work, a specially formulated compound starch-based adsorbent (CSA), which consists of corn, sweet potato and foaming agent, was developed. As an exploratory study, this research was focused on the investigation of CSA for biosorptive dehydration of aqueous ethanol and optimization of conditions for its maximum adsorption under various conditions of vapor feed flow rate, bed temperature and bed height through an orthogonal design method. Furthermore, to investigate the adsorption mechanism, dynamic saturated adsorbance of CSA was studied and characterization of the adsorbent before and after adsorption, as well as after regeneration, was performed.

In the first step, the net retention time and separation factor of ethanol and water were obtained using inverse gas chromatography (IGC), the results indicated that the net retention time of water on CSA is significantly longer than that of ethanol which attested the feasibility of this adsorptive separation purpose and low temperature could be propitious to the process which was consistent with the traditional theory of adsorption.

Then, through a batch adsorption experiment, the optimum adsorption condition was determined under vapor feed flow rate, bed temperature and bed height by orthogonal design method, which indicated that the optimization condition for ethanol was at vapor feed flow rate of 1.1g/min, bed temperature of 83   and bed height of 25cm. The packing density was 445kg/m^3. The residence time is determined by vapor feed flow rate and bed height. The residence time is more influenced than bed temperature. The longer the residence time, the more the output is. But it cannot be ignored that the adsorption selectivity may decreases. Low bed temperature is benefited to the adsorption process which agrees with the result of IGC. Five cycles experiments of adsorption-regeneration were carried out at the optimum adsorption and regeneration condition, the results showed that there is practically no change in the regeneration efficiency of CSA after the cyclic process adsorption–regeneration. It is due to the properly chosen regeneration conditions assuring complete desorption of the water adsorbed without changes in the CSA structure. This unchanged adsorption capacity after a number of cycles demonstrates reusability of the adsorbent.

To study the adsorption mechanism, experiment was conducted to study the dynamic saturated adsorbance of CSA. The water adsorption capacity of CSA has been found to be similar to some molecular sieves, which is about 0.15g/g. Besides, field emission scanning electron microscopy (FESEM) and mercury porosimetry (MP) were applied to investigate the change of surface morphology and microstructure of CSA before and after the adsorption, as well as after regeneration, as shown in Fig.1 and Fig.2. The FESEM images showed that compared with the particles before the adsorption, they became plump and the void spaces became smaller after adsorption, which indicated that the CSA was saturated with water. However, the CSA particles after regeneration became wizened like the nature particles, which demonstrated that almost all the water had been desorbed from CSA. The pore size distribution (PSD) indicated that compared with the PSD curve before adsorption, the peak at 6000nm disappeared and the peak at 17000nm became remarkably smaller after adsorption, which manifested that CSA is mainly composed of macropores which play a major role in the adsorption. However, the peaks at 6000 and 17000nm reappeared after regeneration, and the curve looked exactly the same with that before adsorption, which further demonstrated that CSA after regeneration can be totally regenerated and recovered to the original structure.

These results indicated that the starch-based adsorbent can be considered as an effective and low-cost bio-sorbent to dehydrate alcohol.

349781 Purification of 2-Amino-1-Phenylethanol Enantiomers By a New Technique Combining Distillation and Crystallization

Abstract:

Stripping crystallization (SC) is a new separation technique which combines distillation and crystallization. Basically, SC is operated at a triple-point condition, in which the liquid is simultaneously vaporized and crystallized due to the three-phase equilibrium. In essence, the process is continued until the liquid phase is completely eliminated and only the pure solid crystals remain in the feed. Thus, crystal washing is not required since only pure solid crystals remain in the feed and no impurities are adhered on the crystal surfaces at the end of SC. In the previous research, SC has been successfully applied to separate p-xylene from the mixed xylenes by Shiau et al. (2005, 2008).

Phenylethanolamine, also known as 2-amino-1-phenylethanol (2-APE), is an important basic skeleton in many adrenergic drugs and is often used as ingredients of cold medicines and anorectics. As the biological activity and pharmacokinetics of the enantiomers of a chiral drug can be different, the objective of this research is to apply SC to purify R-2-APE from the chiral 2-APE mixture. The experiments show that no crystallization occurs during SC without seeding due to subcooling of the liquid enantiomers. On the other hand, seeding can eliminate subcooling during SC, leading to the R-2-APE crystals gradually formed preferentially on the seeds. Thus, seeding with magnetic stirring can be successfully employed to purify R-2-APE from the 2-APE enantiomers.

1. Shiau, L. D., Wen, C. C. and Lin, B. S. Separation and purification of p-xylene from the mixture of m-xylene and p-xylene by distillative freezing, *Ind. Eng. Chem. Res*, **44(7)**, 2258-2265, 2005.

2. Shiau, L. D., Wen, C. C. and Lin, B. S. Separation of p-xylene from the multicomponent xylene system by stripping crystallization, *AIChE Journal*, **54(1)**, 337-342, 2008.

350841 Fabrication of Low Cost Insulating Material from Kaolin Clay for Construction Purposes

Abstract:

A new low cost, thermal insulator, construction mineral polymeric material was synthesized from local Saudi Arabia resources using Inorganic Mineral Polymerization technique (MiP). In this technique, the mineral polymer is produced from the following components: clay mineral rich in koalinite, sand, sodium hydroxide solution, and water. Inorganic polymerization experiments are conducted by mixing known amounts of clay minerals and sand in a variable speed stainless-steel mixer for some period of times until a homogenous solid mixture is obtained. Sodium hydroxide solution is added to the mixture and the polymerization reaction takes place at different mixing speeds and times. The produced mixture is divided into a number of samples of known weights. Each sample is moulded in a stainless-steel cylinder at a medium pressure. Then the moulded samples are cured in an oven for 24 hours. The curing temperature was 80°C. The mineral polymerization conditions such as the ratio of clay to sand filler, particle size of the clay, particle size of the sand, concentration and amount of sodium hydroxide, speed and time of mixing were optimized to obtain stable and hard mineral polymer. Compressive strength for different specimens of fabricated mineral polymers was tested under dry and wet conditions. Values of compressive strength with values ranging from 11.5 MPa to 27.9 MPa were obtained.

351141 Thermodynamic Study of Binary PAH (Anthracene + Phenanthrene) Solid Mixtures

Abstract:

Polycyclic aromatic hydrocarbons (PAHs) are common components of many materials, such as petroleum and various types of tars. They are generally present in mixtures, occurring both naturally and as byproducts of fuel processing operations. It is important to understand the thermodynamic properties of such mixtures in order to better understand and predict their behavior (i.e., fate and transport) in the environment and in industrial operations. The Knudsen effusion technique was combined with gas chromatography and differential scanning calorimetry to measure the vapor pressure, composition, and phase behavior of binary anthracene and phenanthrene mixtures at near ambient temperatures. Mixtures of anthracene and phenanthrene exhibited non-ideal mixture behavior: In mixtures with less than 80% phenanthrene, this more volatile component preferentially sublimed and the vapor pressures decreased as the mixtures became enriched in anthracene. The vapor pressure data do not follow Raoult's law (that is, ideal mixture) predictions. Mixtures with high phenanthrene content (> 80%) showed azeotrope like behavior, in that anthracene and phenanthrene sublimed at nearly constant vapor pressure and molecular ratio. These results confirm that the behavior of few-component PAH mixtures does not follow ideal mixture models. Hence, as tars and petroleum (and potentially other organic compound-containing mixtures) weather to few component mixtures over time, complicated, non-ideal phase behavior might be observed.

351160 How Confident Are You That Your Major Accident Risks Are Under Control?

Abstract:

Major accidents in the process industries including those at Texas City , Buncefield UK and in the Gulf of Mexico have highlighted the critical role played by senior managers and executives. Effective leadership is essential if companies are to develop a positive safety culture that remains constantly vigilant towards the risk of a major process safety incident. In the UK the HSE has made leadership a key priority for the high hazard industries and there is a cross-industry initiative to develop and deliver training focussed towards board members responsible for process safety performance.

This paper explores the leadership failings that contributed to recent major accidents, and the cost to both the company reputation and the individuals involved. Key principles are discussed that can help improve leadership in the major hazard industries. These include; attitudes and decisions of senior manager affecting the investment strategy and the safety culture of the organisation, reinforcing the importance of safety by personal example, thorough understanding of major accident hazards and key risk control systems, investigating process safety incident and near misses to find the underlying causes, developing world class safety management systems, and identifying weaknesses in these systems using targeted performance indicators.

The author bases this paper on the development of a training standard for 'Process Safety Leadership' involving a cross-industry 'expert panel'. Training of senior executives in compliance with this standard has been carried out from late 2011 and feedback from these courses and the key requirements for effective leadership will be discussed.

351236 Sil Determination: Shortcomings with the Use of LOPA

Abstract:

Proposed Topic Area: Managing Risks

For a number of years, industry has been trying to apply the principles outlined in IEC 61508. In the process sector, this has resulted in a focus on the requirements of IEC 61511 (2003) and ANSI/ISA 84.00.01 (2004). These are risk based standards. They involve the identification of potential hazardous events, the causes that could lead to each event and the necessary risk reduction for each event. Where the risk reduction involves an instrumented function (e.g. a trip or an interlock), the required performance of the instrumented function can described in terms of the average probability of failure on demand. In the standards, the performance for such instrumented functions is assigned to one of four Safety Integrity Levels. The assessment of what performance level is needed in respect to a specific hazardous event is referred to as SIL Determination. In the field of SIL Determination, IEC 61511 provides examples of a number of different approaches – these are shown in IEC 61511 Part 3. These include Event Trees, Risk Graphs, Safety Matrices, Fault Tree Analysis and Layer of Protection Analysis (LOPA).

In the first few years, following the publication of IEC 61508 Edition 1 in 2000, industry generally embraced the use of Risk Graphs as the method of choice. However, over the subsequent few years, many companies using this method found that it was (a) giving higher safety integrity level targets than really needed and (b) was less flexible than other methods and consequently more difficult to use in many real situations. This has resulted in a general move to other methods and in particular to the use of Layer of Protection Analysis. Indeed, LOPA may now be the predominant "Method of Choice" for SIL Determination. As a method, LOPA gives the impression of simplicity and appears quite straightforward. It almost invites anyone with a minimum of understanding of the method to be a practitioner. However, there are a number of shortcomings in the use of the LOPA method that have become evident through its use across the globe.

This paper will review the application of LOPA and provide insights into what may be seen as the proper use of the method and its core principles. It will also highlight various shortcomings in use that have been found during expert reviews carried out of a number of existing studies. The paper will be of benefit both to those who are new to the LOPA method, but also of value to those who have been using the method for some time.

Shale Oil and Gas – Wellhead to Fuel

Dr. Lawrence Kremer - Baker Hughes
AIChE - Hilton New Orleans Riverside
April 1, 2014

14 AIChE
Spring Meeting, New Orleans, LA

BAKER HUGHES

Outline Unconventional Shale Oil

- Properties of shale oils
- Problems
  - Transportation
  - Refining

14 AIChE
Spring Meeting, New Orleans, LA

BAKER HUGHES

General Characteristics of Shale Oils

- API gravity typically varies 30-55+
- Very paraffinic
- Low sulfur
- Can have hydrogen sulfide and mercaptans
- Low TAN
- Very low asphaltenes
- Light crude with very little residual material
- Many samples show great quality variation

Factor	Bakken	WCS
API	39.7	21.6
S%	0.25	3.44
Light ends	2.6%	
LSR	3.5%	3%
Light Nap	12.4	3.0
Heavy Nap	9.7	5.0
Jet	10.5	10.0
LFO	27.6	12.0
Gasoil	27.2	37.0
VacTar	6.7	29.8

14AIChE Spring Meeting, New Orleans, LA

BAKER HUGHES

Shale Oil Property Comparison

Parameter	Eagle Ford	Bakken
API	52	40.8
TAN (g KOH/g)	< 0.05	0.09
Sulfur (wt %)	<0.2	0.304
Asphaltene (wt %)	0.1	0.41
Resin (wt %)	1.6	4.95
Filterable Solids (PTB)	225	76

14AIChE Spring Meeting, New Orleans, LA

BAKER HUGHES

Utica Shale Crude - Three Phase

14 AIChE
Spring Meeting, New Orleans, LA

BAKER HUGHES

Eagle Ford Crude Samples

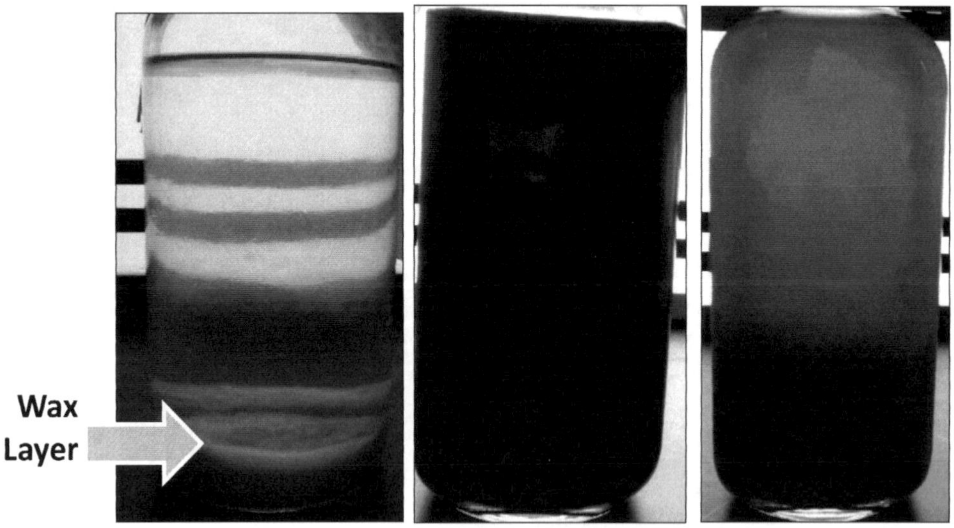

14 AIChE
Spring Meeting, New Orleans, LA

BAKER HUGHES

Analytical Breakdown

Value	Yellow	Red	Black
API	55.0	44.6	52.3
TAN	<0.05	0.07	<0.05
Sulfur	<0.20%	<0.2%	<0.2%
Na	1.0 ppm	1.6 ppm	1.6 ppm
K	0.3 ppm	0.4 ppm	0.5 ppm
Mg	3.4 ppm	2.9 ppm	3.0 ppm
Ca	2.6 ppm	2.8 ppm	3.8 ppm
Asphaltene	0%	0%	0.1%
Resin	0.5%	3.2%	1.6%
Filterable solids	176 PTB	295 PTB	225 PTB
P	5.3 ppm	4.6 ppm	5.0 ppm

14AIChE Spring Meeting, New Orleans, LA

BAKER HUGHES

Micrograph of Sludge

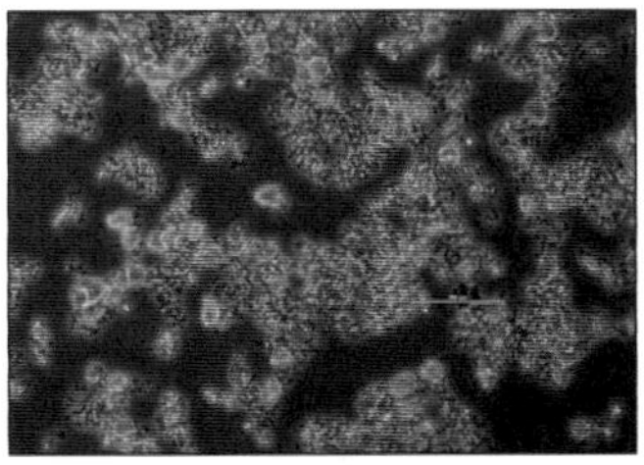

100 X

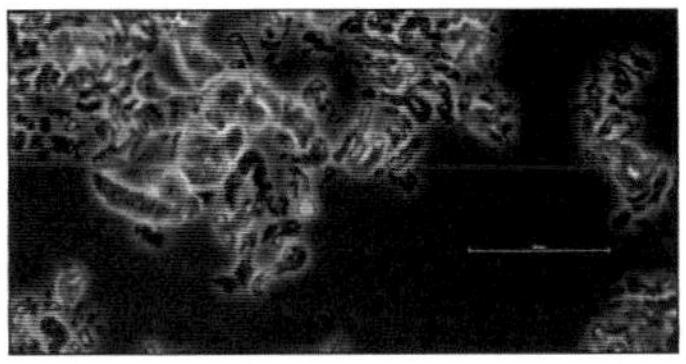

400 X

14AIChE Spring Meeting, New Orleans, LA

BAKER HUGHES

Paraffin Analysis of Sample

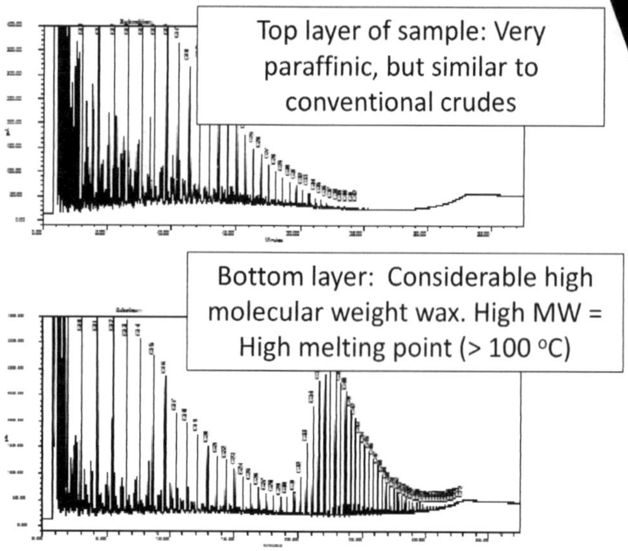

14AIChE
Spring Meeting, New Orleans, LA

BAKER HUGHES

Transport of Shale Production

- Infrastructure does not exist to meet needs
  - Most shale oil moved by rail or trucks
  - Limited pipelines expanding to meet transportation needs
- Unit trains to East Coast, West Coast, and Gulf Coast
- Problems:
  - Tankers and barges with heavy wax deposits ROB
  - Tight emulsions that look like oil or white condensate but high BS&W
  - Cars not uniformly treated for H_2S
  - Corrosion in rail cars

14AIChE
Spring Meeting, New Orleans, LA

BAKER HUGHES

Viscosity of ROB Before Treatment

ROB from untreated compartment

14AIChE®
Spring Meeting, New Orleans, LA

BAKER HUGHES

Eagle Ford Shale Oil Wax Dispersion Testing

Several products were effective in dispersing the paraffin wax in an Eagle Ford shale oil sample at ambient temperature for over a month.

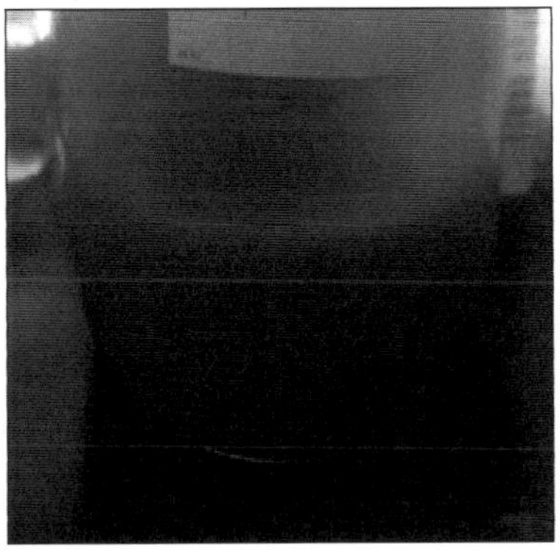

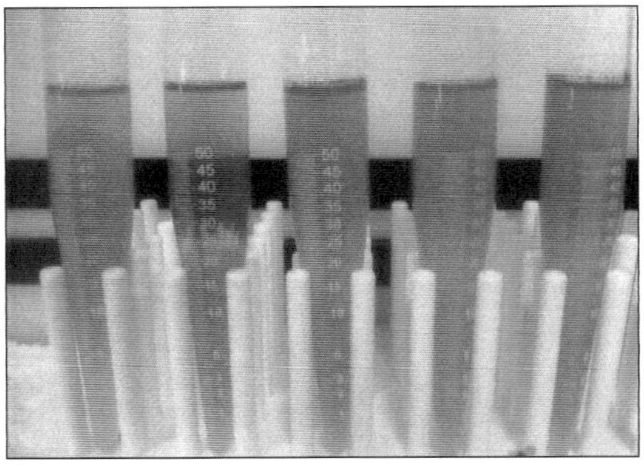

14AIChE®
Spring Meeting, New Orleans, LA

BAKER HUGHES

Product 'A' Effect on Wax Deposition (Utica)

Blank 200 ppm 400 ppm

14AIChE
Spring Meeting, New Orleans, LA

BAKER HUGHES

Refinery Issues

- Crude Blending
- Fouling
- Corrosion
- Product quality

14AIChE
Spring Meeting, New Orleans, LA

BAKER HUGHES

Crude Unit Fouling

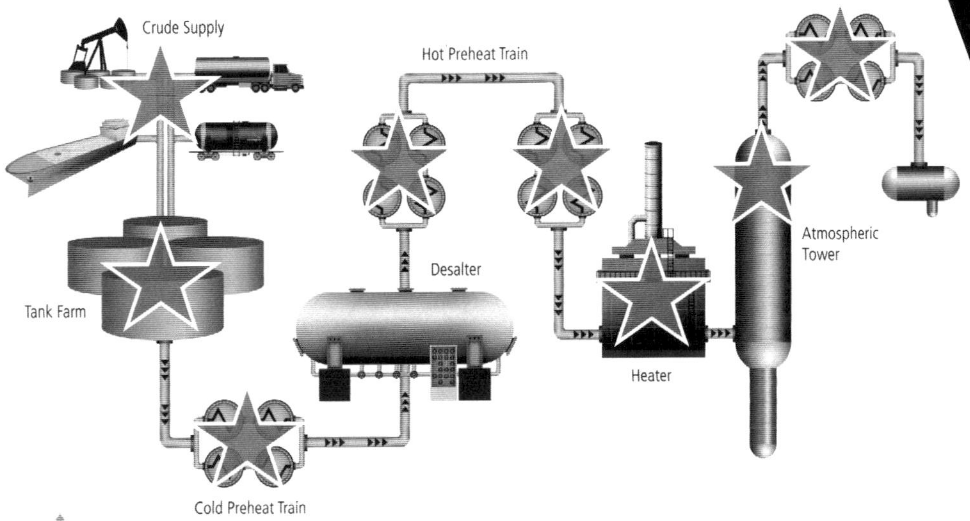

★ = Susceptible to Fouling

14 AIChE Spring Meeting, New Orleans, LA

BAKER HUGHES

Asphaltene Precipitation

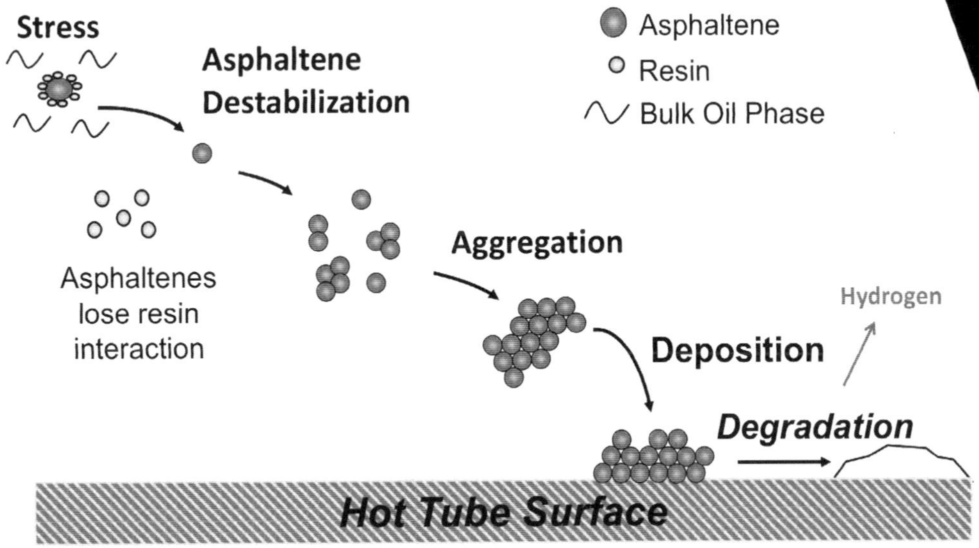

14 AIChE Spring Meeting, New Orleans, LA

BAKER HUGHES

Asphaltene Stability Index Test

Field ASIT services™ technology measures the onset of the flocculation of the asphaltenes with high accuracy by inducing the asphaltene precipitation via titration with a paraffinic solvent.

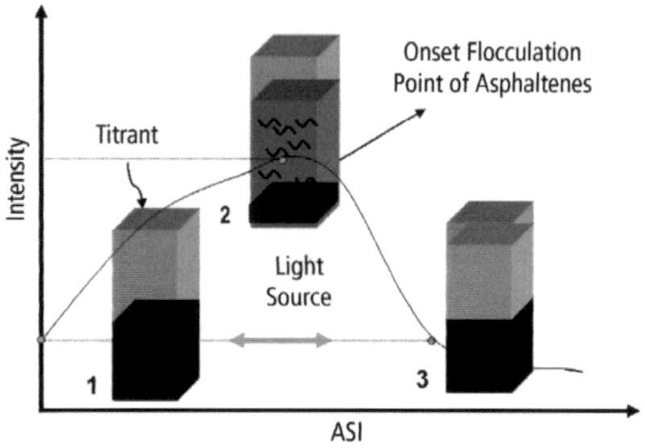

14AIChE Spring Meeting, New Orleans, LA

Field ASIT services is a trademark of Baker Hughes Incorporated

BAKER HUGHES

ASI Shows Eagle Ford Can Destabilize Crude

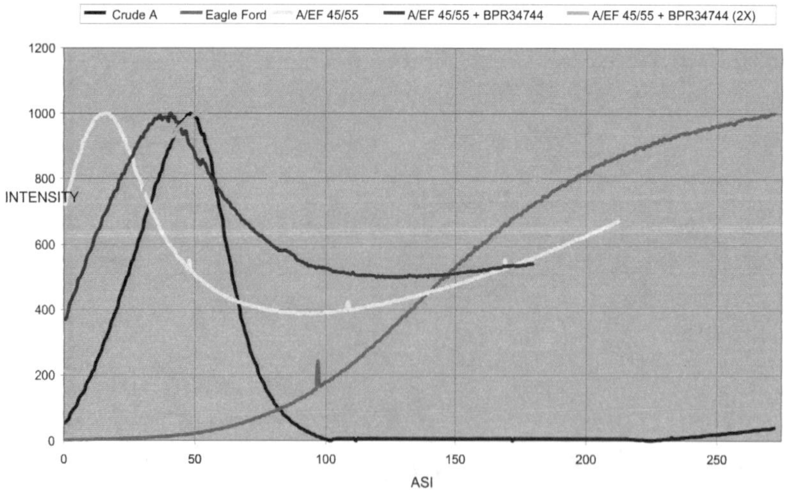

14AIChE Spring Meeting, New Orleans, LA

BAKER HUGHES

Tower Tray Fouling - Phosphorus

- Seen by refineries running Bakken, Eagle Ford, and WTI
- Operationally manifested as increasing tower dP
  - Typically in the kero/jet zone
- Tower inspection shows deposits on trays in kero/jet zone
  - Some deposition may also show a few tray levels above and/or below kero/jet zone
  - Deposit scale typically tightly bonded to tray surface restricting flow and valve function

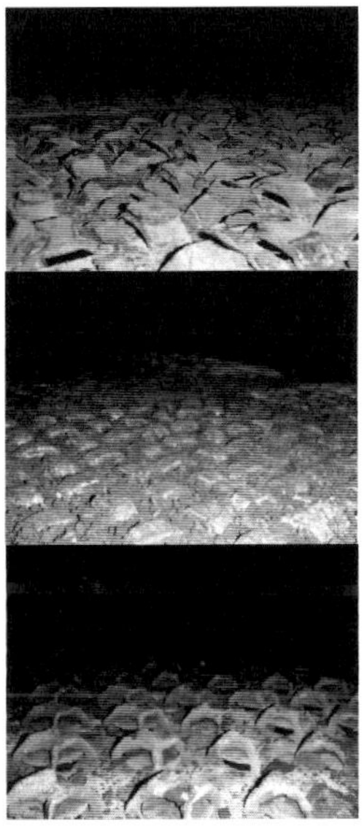

Shale Oil Impact – MEA Contamination

Corrosion and Fouling

- Shale oils transported by truck, rail, barge.
- Many shipments are treated with H_2S scavengers:

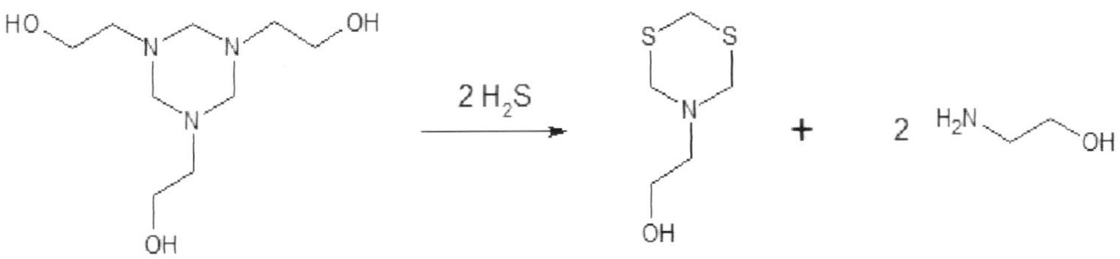

- Potential Impacts:
  - H_2S scavengers leave byproduct amine in crude oil (MEA, MA).
  - Tramp amines distill to OH system, create amine-HCl salts:
  - Corrosive to system metallurgy.

Impact of Tramp MEA on OH Water

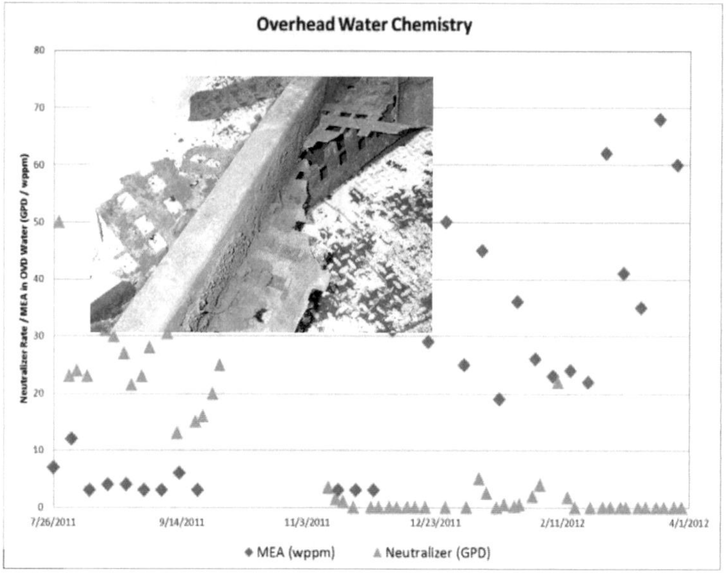

14AIChE®
Spring Meeting, New Orleans, LA

BAKER HUGHES

Salt Risk w/MEA Concentration

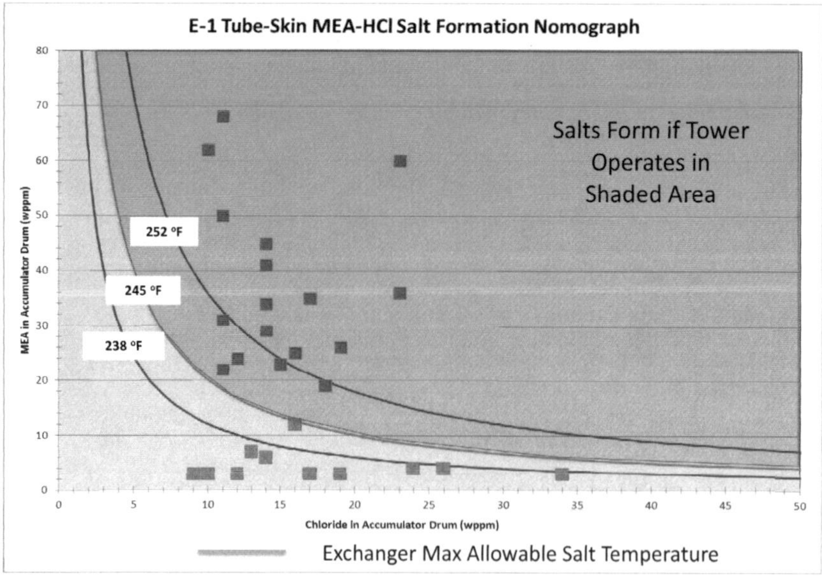

14AIChE®
Spring Meeting, New Orleans, LA

BAKER HUGHES

Shale Oil – Refinery Process Concerns and Monitoring

BAKER HUGHES

CHARACTERISTIC OF SHALE OIL	CONCERN IN REFINERY	PROCESS DATA	SAMPLES AND ANALYSES Samples	Analysis	OUTPUT DATA	FREQUENCY	BHI PROGRAMS
Crystalline Wax	Tank bottom sludge	• Tank Sludge Depth	• Tank bottom	• Visual sludge • GC analysis	• Tank Sludge Depth • Crystal Wax Presence	• Monthly	• XERIC™ Pretreatment
	Cold Preheat Fouling	• Desalter Temperature			• Temp vs. Time	• Update 2x/mo.	• LIFESPAN™ Wax Dispersants
	Desalter emulsions	• Emulsion layer growth	• Triline emulsions	• Visual wax • GC analysis	• Crystal Wax Presence	• After Shale Oil begins • When emulsion grows	• XERIC Demulsifiers
Asphaltene Stability in Blends	Desalter emulsions	• Emulsion layer growth	• Triline emulsions	• Precipitated Asphaltenes	• Precipitated Asphaltenes	• After Shale Oil beings • When emulsion grows	• XERIC Asphaltene Stabilizers • Field ASIT services™
	ColdPreheat Fouling	• Desalter temperature	• Crude blend	• ASIT	• Temp vs. Time • ASIT #	• Pre-Shale Oil • Update 2x/mo.	• LIFESPAN Antifoulants • Field ASIT services
	Hot Preheat Fouling	• Furnace inlet temp • Furnace tube temps • Furnace fuel usage • Preheat DP	• Crude blend	• ASIT	• Temp vs. Time • Fuel use vs. Time • ASIT #	• Pre-Shale Oil • Update 2x/mo.	• LIFESPAN Antifoulants • MILESTONE™ Furnace Fouling Control • Field ASIT services
Paraffinic	Furnace Coking	• Furnace tube temps • Furnace fuel usage	• Crude blend	• HLPS	• Coking tendency	• Pre-Shale Oil • W/Shale Oil	• MILESTONE Furnace Fouling Control
Tramp MEA	Tower/OH corrosion (Early Warning)	• OH Neutralizer rate • EXCAL usage rate	• OH water • Desalter brine	• OH pH • Brine pH	• Early Sign of Amine • Early sign of Amine	• Pre-Shale Oil • W/Shale Oil	• TOPGUARD™ Overhead Corrosion Control • EXCALIBUR™ Contaminant Removal Technology
	Tower/OH corrosion (Quantitative)	• Inputs for CRM	• OH water • Desalter ww • Desalter brine	• IC Amines	• Confirm MEA in Crude • OH MEA ppm • Salt Risk Tower • Salt Risk OH System	• Pre-Shale Oil • W/Shale Oil • Update 2x/mo.	• TOPGUARD Corrosion Control • EXCALIBUR Technology • TOPGUARD Corrosion Risk Monitor (CRM)
Phosphorus	Tower P Fouling	• Tower DP			• Tower DP Increase	• 2x/mo.	• LIFESPAN Antifoulants Dispersants
			• Crude	• ICP for P	• P in Crude	• 2x/mo.	
			• Kerosene	• Fill Solids • XRF Filter P • ICP for P	• Incr. Kero Solids • P in Solids • P in Kerosene	• 2x/mo.	
			• Kerosene • Diesel	• ASTM D86	• Incr. in Diesel IPB – Kero EP overlap	• 2x/mo.	

EXCALIBUR, FIELD ASIT SERVICES, LIFESPAN, MILESTONE, TOPGUARD and XERIC are trademarks of Baker Hughes Incorporated.

An Experimental Investigation of Settling Velocity of Spherical Particle in Newtonian Fluid using Particle Image Shadowgraphy

Shivam Shahi, Martha Wenxin Han and Ergun Kuru

University of Alberta, Edmonton, AB, Canada

Abstract

The settling velocity is a fundamental requirement and key variable for modeling sedimentation processes and particle transport. The Particle Image Shadowgraphy (PIS) experimental technique used here is unique in its kind and very efficient in measuring the size, shape, and settling velocity of the particles, simultaneously. Experiments are conducted using different sizes of glass and plastic spheres (0.5-2 mm) with five different Newtonian fluids. Mixtures of water and different concentrations of glycerol (10-40% by volume) have been used as the fluid medium in the experiments. Different models to predict the settling velocity present in the literature have been critically analyzed and based on the experimental results from the shadowgraph a new dimensionless model for predicting settling velocity for a wider range of particle size and flow regime has been developed.

Introduction

The earliest literature to determine the relationship between settling velocity and particle moving through fluid could be traced back to the Stokes' Law (1851), which was derived by equating the effective weight of a spherical particle to the viscous resistance. Stokes' Law applies only for small rigid spheres settling within an infinite and viscous fluid column. Since early 20th century, numerous studies have been conducted and several empirical formulae were established to estimate the setting velocity of a particle. Rubey (1933) combined Stokes' Law and the Impact Law into a general equation, which considers not only the viscous resistance but also the fluid impact. Gibbs et al. (1971) derived the empirical equation to estimate the relationship between settling velocity and grains of sphere size. Peden and Luo (1987) proposed the drag coefficient correlations for spheres within a limited particle Reynold's number range, which when compared with the experimental results gave an error of ±16%. Mordant and Pinton (2000) experimentally investigated the motion, physics and factors governing the setting of solid spheres using acoustic models and signal processing Later on, Brown and Lawler (2003) studied different settling velocity correlations and experiments available in the literature and suggested two different correlations for spheres.

In the case of non-spherical particles, there have been several reports in literature that examine the settling velocities through analytical, empirical or phenomenological models (Hallermeier (1981), Cheng (1997), Dietrich (1982), Ahrens (2000)). However, in most of them, the settling velocities were developed based on settling of spheres. Thus, there emerges a need to accurately estimate and understand the settling of spheres in real-world industrial applications. Although the factors influencing the settling velocity are well known, there is always a need for simulating the functional relationships among settling velocities, particles and the influencing parameters more accurately through modern experimental techniques.

This work is an attempt to develop a more accurate empirical model for settling velocity of spheres based on an advanced laser based image processing technique, PIS, for measuring size, shape, and settling velocity. Experiments are conducted using different sizes of glass and plastic spheres (0.5-2 mm) with five different Newtonian fluids. Mixtures of water and different concentrations of glycerol (10-40% by volume) have been used as the fluid medium in the experiments. Further, based on measured experimental values a dimensionless model of particle settling velocity has been has been developed and compared with the predictions of the existing models in literature for spheres in Newtonian fluid.

Experimental

The PIS technique used for visualizing settling of fine particles in fluids, is based on high resolution imaging with pulsed backlight illumination. The general principle associated with this phenomenon is that a shadow is casted whenever there is a significant change in the densities of medium through which the light is passed. This technique generates a series of shadow images from the focal plane of the camera to be projected on screen; allowing particles to be visualized. Most light rays are refracted and the ones un-deflected will appear as dark marks captured by the instrument (Davis 8.0 Manual). This technique is independent of the shape and physical properties (e.g. density, size, transparency) of the particles. It allows for the investigation of sizes down to 5μm and freeze motions of more than 100 m/s when using the appropriate imaging system and light source (Davis 8.0 Manual). The experimental setup designed for measuring the settling velocity using PIS is shown in Figure 1.

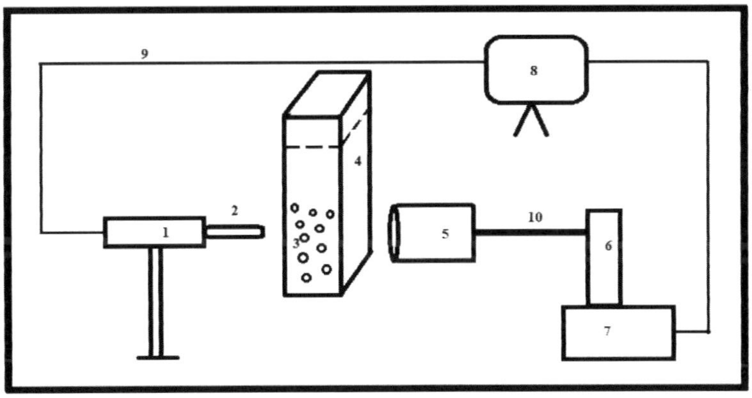

Figure 1: Schematic of the experimental setup for measuring settling velocity using PIS [1. Lavision Image Intense Camera 2. 12 X Navitar Lens 3. Particles 4. Plexi Glass container with Fluid 5. Diffuser 6. Lavision Illumination High Efficiency Diffuser 7.New Wave Research Laser Solo III 15Hz] 8. Lavision Davis 8.0 9. Connecting wire 10. Optical Fibre Cable

A double-pulse laser combined with a double frame camera allows investigating size dependent velocities (Davis 8.0 Manual). This camera enables the use of adjustable image recording frequency. The image processing in Davis8.0 software involves the location of particles in an area of interest and subsequent analysis of their position at each frame. The displacement of particles between two consecutive frames is used to calculate the velocity of the particle as shown in Figure 2.

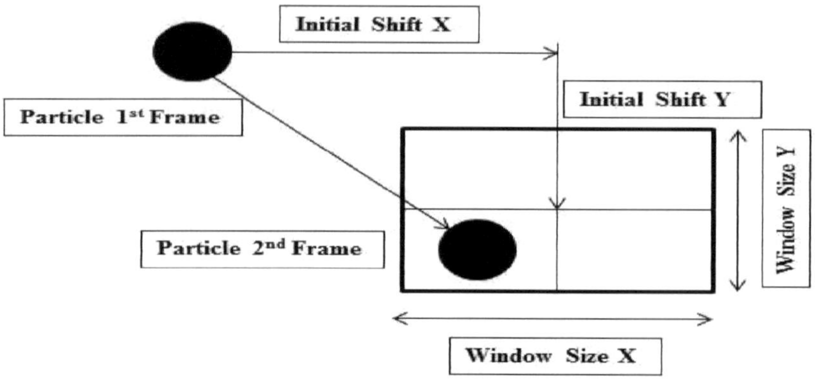

Figure 2: Illustration of interrogation window used to determine velocities of particle

Experiments have been carried out with four different glass spheres (Corposular's Glass Spacers Millibeads) and two different sizes of plastics beads. All the experiments were performed at room temperature. The rheological properties of the fluids were measured using Cannon-Fenske viscometer. Experiments begin with the calibration of the camera, which is done using a rectangular plate with printed circle of size (0.8 mm) spaced at 1.5 mm distance from each other. Once the particle start descending, the double frame camera starts capturing the images at a particular frequency (usually 500-3000 ms), which depends on the size of the particle. The recorded images are processed using Davis 8.0 software which detects the particle based on the intensity difference and calculates the settling velocity.

Validation of Experimental Measurement

The validation of the particle size measurements using PIS technique has been carried out by comparing the measured size and settling velocities of four different glass spheres (Corposular's Glass Spacers Millibeads) with the data provided by the manufacturers. The comparative results of the actual and experimentally measured diameters of glass beads are shown in Table 1.

Table 1: Measured and Actual Diameter of Glass Spheres

Diameter (Manufacturer Data) (mm)	Diameter (Measured) (mm)
0.71±0.02	0.71

2±.04	2.002
1.18±0.02	1.175
1.5±0.03	1.52

In addition, measured settling velocity values from shadowgraph experiments were validated by first determining the drag coefficient using the measured values of settling velocities of spheres in water by using the equation 1 and introducing these values into the universal drag coefficient versus particle Reynolds number plot for Newtonian fluid as shown in Figure 3. It was found that the measured drag coefficients were within acceptable limit of 5% to the theoretical values.

$$C_D = \frac{4gD(\rho_s-\rho_f)}{3\rho_f V_s} \text{............} (1)$$

Where, C_D is drag coefficient, ρ_s is density of solid, ρ_f is density of fluid, V_s is settling velocity, g is acceleration due to gravity, and D is particle diameter

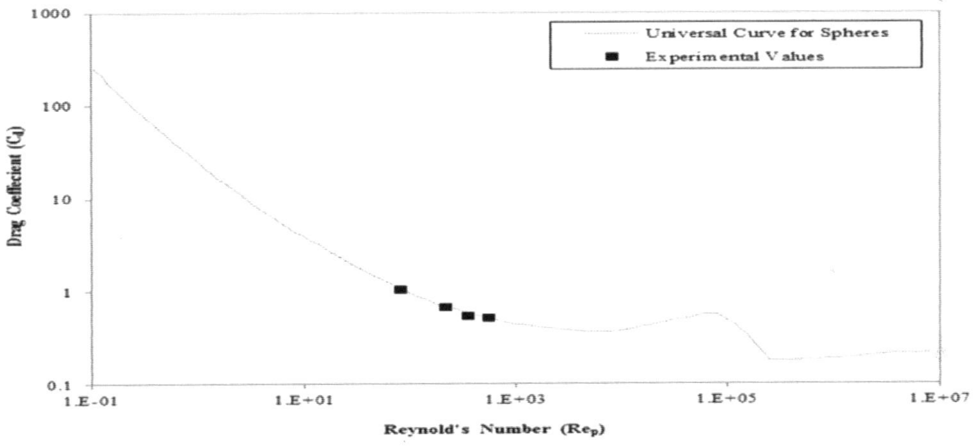

Figure 3: Shows the experimental values on C_D vs Re_p Universal Plot

Fluid and Particle Characteristics

Three different classes of Cannon Fenske Viscometers (SR25 135, SR100 461, and SR150 798) were used for measuring the kinematic viscosity of glycerol water mixture with different concentration (0%-40% by volume) of glycerol. The kinematic viscosity values of five different water-glycerol mixtures are shown in figure 4. The densities of the fluids were determined by using direct mass and volume measurements. A 10 ml measuring jar and highly precise weighing machine was used for these measurements. The measured density and kinematic viscosity values are given in figure 4.

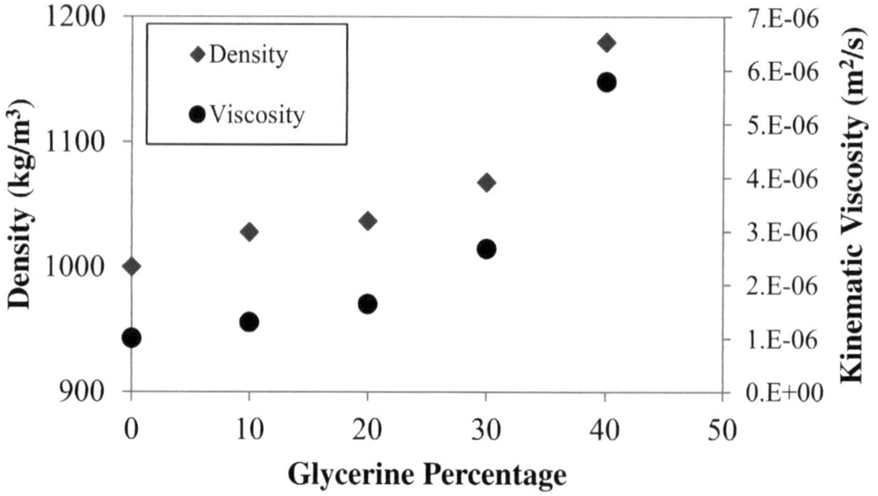

Figure 4: Shows the kinematic viscosity and density values for water – glycerol mixture

The specific gravity of the glass spheres were found to be 2.523 which, compares well with manufacturer's value of 2.510. Considering the possible errors in measurements, we have decided to use the manufacturer's data in our analyses.

Results and Discussion

Measured Settling Velocities

Experiments were performed using four different sizes of glass spheres (0.71, 1.18, 1.5 and 2.0 mm) in five different glycerol-water mixture (0-40% glycerol by volume in water). Each fluid and sphere experiments were triplicated to verify repeatability better accuracy for the measurement. The average settling velocities of a sphere in each particular fluid are reported in Figure 5. It can be seen that the settling velocity increases steadily with the particle diameter. In addition, the settling velocity is inversely proportional to the viscosity of the solution for all particle sizes. This is consistent with the physics of settling particles and many other experimental results available in literature (Gibbs et al. (1971), Brown (2003), and Mordant (2000)). Moreover, the settling velocity obtained from the PIS experiments were compared with the predictions of a few well accepted models. The models include, Rubey (1933), Ahrens (200), Darby (2009), Hallermeir (1981), Julien (1995),Soulsby (1997), Van Rijn (1989), Zanke(1977), Sha (1954), Cheng (1997) for predicting settling velocity of a particle in Newtonian fluid. The model predictions and the experimental results are summarized in Figure 6(a). Furthermore, the same model predictions were compared with some of the experimental data from Gibbs (1971) and the results are shown in Figure 6(b). The analyses of comparative results show that all the considered models underestimate the settling velocities. One can observe a similar trend in both the plots, the difference in the predicted and experimental values are increasing as the diameter of the particles increase. In other words, as the size of the particles reduces, accuracy of these model predictions improves. It can also be noticed that no two models give the same results for particle's settling velocity

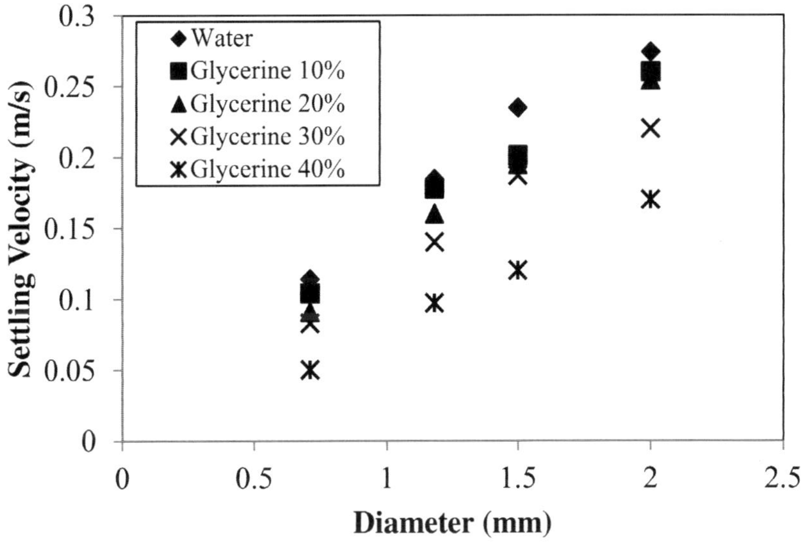

Figure 5: Measured settling velocity of glass beads in different water-glycerol mixture

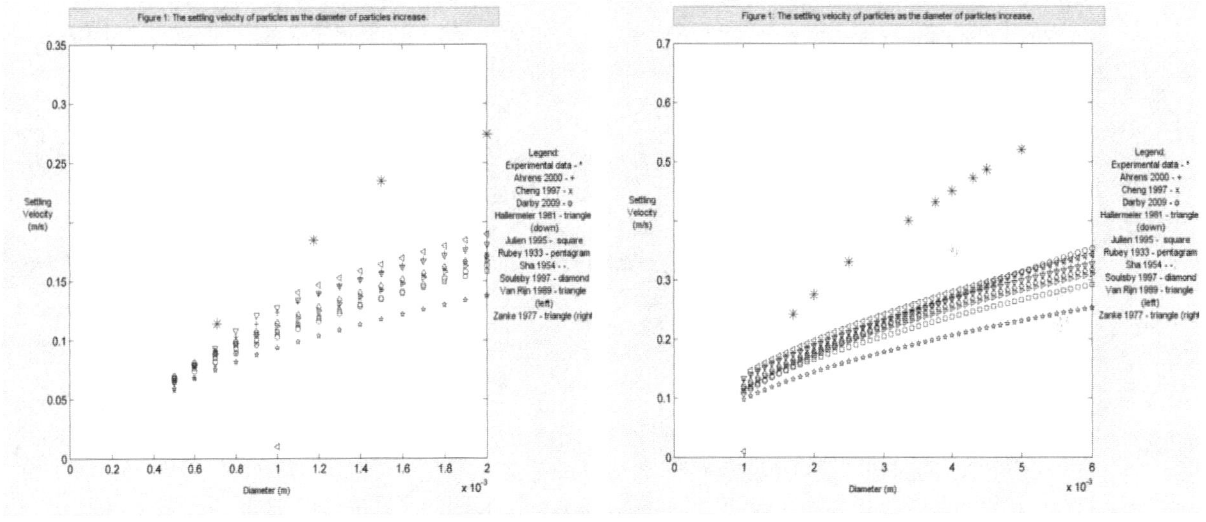

(a) **(b)**

Figure 6: **(a)** Shows the predicting capability of available models compared with shadowgraph's result b) Shows the predicting capability of available models with Gibbs experimental result

Development of the Settling Velocity Model

The empirical model developed here is based on the dimensionless diameter and velocity. In the current analysis, extended version of Hallermeir's number (1981) with Archimedes Buoyancy

Index is used. Similar parameters have also been used by Cheng (1997) and Dietrich (1982). The two dimensionless numbers used in this study are given by the equations 2 and 3.

$$D_d = (\rho_s - \rho_f) * D^3 * \frac{g}{\rho_f * v^2} \quad \text{......... (2)}$$

$$V_d = V^3 * \rho_f/((\rho_s - \rho_f) * g * v \quad \text{......... (3)}$$

The measured values of settling velocities from the shadowgraphy experiments shown in Figure 5were fit to a power law model (See equation 4). The regression coefficient value comes out to be 0.991 for the experimental data. Due to the limited range of experimental data the equation 4 predicts the settling velocity values very efficiently in 1.18 E+02 < D_d < 1.2 E+06. This is a major limitation for the empirical models in general.

$$V_d = 0.2269\mathrm{D}_\mathrm{d}^{0.7497} \quad \text{............. (4)}$$

When analyzed and compared with the other experimental values from the literature (Gibbs et al. (1971), Brown (2003), and Mordant (2000)) on a dimensionless scale, all the values are found to be lying on the same curve as shown in Figure 7. This not only verifies the consistency with the dimensionless number but also the accuracy of the results obtained. To have greater scope for the model, experimental results from (Gibbs et al. (1971), Brown (2003), and Mordant (2000)) were also included in the regression fit to yield a regression coefficient (R^2) of 0.9991.

$$V_d = 0.404D_d^{0.695} \quad \text{........... (5)}$$

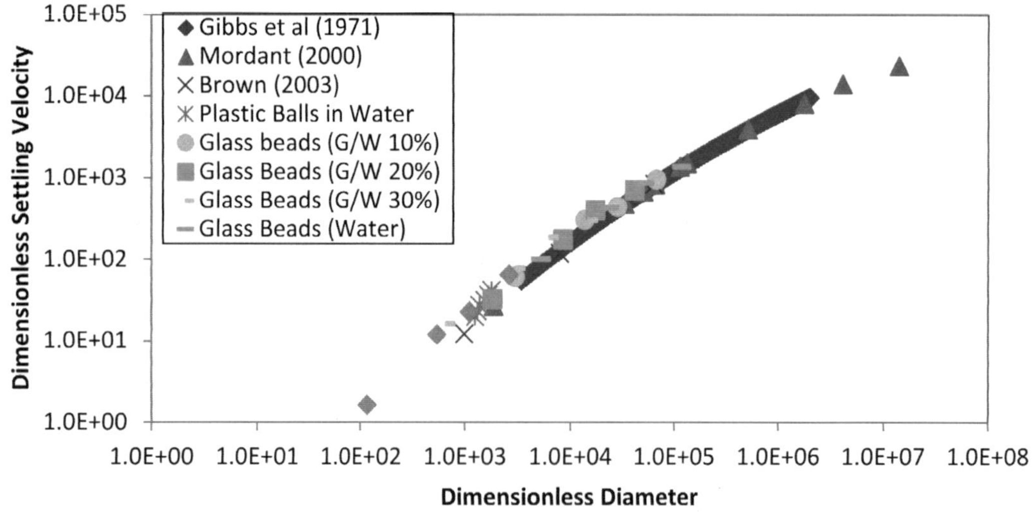

Figure 7: Comparison of Experimental Data (Shadowgraph) and Literature data on dimensionless scale

Conclusions

A new experimental technique for measuring the settling velocity, size, and shape of the particle has been developed. The measurement technique is based on the double frame image processing and is found to be very accurate in terms of size, shape, and velocity measurements. The experimental technique used in this work also describes a new application of shadowgraphy technique in fluid particle systems. Different settling velocity predicting models available from the literature have been analyzed. It was found that most of the models underestimate the settling velocity values. In a quest for more accurate predicting model, a new empirical model has been developed using dimensionless diameter and settling velocity. Based on the value of dimensionless diameter value one can either use equation 4 or equation 5 correlation to get the dimensionless settling velocity.

References

1. "Product Manual: ParticleMaster Shadow," Item Number: 1105032, Lavision GmBh Germany.
2. Ahrens, J.P. (2000), "A Fall-Velocity Equation," Journal of Waterway, Port, Coastal, and Ocean Engineering, 126, pp.99-102.
3. Brown,P. Phillip.,and Lawler, F. Desmond. (2003), "Sphere Drag and Settling Velocity Revisited," Journal of Environmental Engineering, 129, pp.222-231.
4. Cheng, N.S.(1997), " Simplified Settling Velocity Formula for Sediment Particle," Journal of Hydraulic Engineering, 123(2), pp.149-152.
5. Dietrich, E William. (1982), "Settling velocity of Natural Particles," Water Resources Research, 18(6), pp.1615-1626.
6. Gibbs, R.J., Matthews, M.D. and Link, D.A. (1971), " The relationship between sphere size and settling velocity," Journal of Sedimentary Petrology, 41 (1), pp. 7-18.
7. Hallermeier, R.J. (1981), "Terminal Settling Velocity of Commonly Occurring Sand Grains," Sedimentology, 28(6), pp.859-865.
8. Julien, Y.P. (1995), "Erosion and Sedimentation," in Cambridge University Press.
9. Mordant, N. and Pinton, J. F. (2000), "Velocity measurement of a settling sphere," The European Physical Journal B, 18, pp. 343–352.
10. Peden, M James., and Luo, Yuejim. (1987), "Settling velocity of variously shaped particles in drilling and fracturing fluid," SPE Drilling Engineering.
11. Rubey, W.(1933), "Settling velocities of gravel, sand, and silt particles," American Journal of Science, 225, pp.117-119.
12. Sadat-Helbar, S.M., Amiri-Tokaldany, E., Darby, S., and Shafaie, A. (2009), "Fall Velocity of Sediment Particles," in IASME/WSEAS International Conference on Water Resources, Hydraulics and Hydrology, 39-45.
13. Sha, Y.Q. (1956), "Basic principles of Sediment Transport," Journal of Sediment Research, 1 (2), pp. 1-54.
14. Soulsby, R.L. (1997), "Dynamics of Marine Sands," in Thomas Telford, London.
15. Stokes, G. G. (1851), "On the effect of the internal friction on the notion of pendulums," Cambridge Philo. Trans., 9 (2), pp. 8-106.
16. Van Rijn, L.C. (1989), "Handbook: Sediment Transport by Currents and Waves," Rep. No. H 461, Delft Hydraulics, Delft.
17. Zanke, U. (1977), "Berechnung der sinkgeschwindigkeiten von sedimenten," Mitt. Des Franzius-Instituts fuer wasserbau, Heft 46, Seite 243, Technical University Hannover.

350459 The Influence of Particle Humidity on Separation Efficiency for an Axial Cyclone Separator

Abstract:

This study aimed at evaluating the influence of particle humidity on separation efficiency for an axial cyclone separator. The collection and grade efficiencies of an axial cyclone separator were measured by on-line method for inlet velocities, particle concentration and particle relative humidity in the ranges of 12-18 m/s, 30-500 mg/m^3 and 8-30‰, respectively. The experimental results showed that the volume proportion of the inlet particles with bigger particle diameter increased with the increasing of particle relative humidity, because the liquid bridge forces among particles were strengthened with the increasing of particle relative humidity, which caused to the particles easily aggregates to bigger ones. If the inlet velocity and particle relative humidity were constant, the improvements on collection and grade efficiencies were obviously with the increasing of particle concentration because of particle aggregation. However, if the inlet velocity and particle concentration were constant, the grade efficiencies did not always increased with the increasing of particle relative humidity. Furthermore, the grade efficiency curves for different particle relative humidity crossed at the particle diameter approximately 10μm, the grade efficiency would be moved up with the increasing of particle relative humidity when the particle diameter bigger than 10 μm, and the trends of the curves were contrast when the particle diameter lower than 10 μm. Because the liquid bridges among particle were easy to be broken by centrifugal force, the higher the particle relative humidity got, the more fine particles escaped from the cyclone separator outlet.

348400 Vibrating Electrical Elevator and Reactor: Universal Continuous Reactor for Thermal Processing

Abstract:

Food thermal processing is widely used for pasteurization as an eco-friendly alternative to chemical treatments. Moreover, the constant evolution of costumer tastes and needs give an additional interest of thermal processing for value-added food (e.g. cereal roasting). Several technologies have been developed for different kind of application using hot air, steam, hot water, hot oil or microwaves [1]. This paper presents a simple and universal food processing technology, based on a direct contact with a hot tube.

REVTECH has developed and supplied industrial solutions for heat treatment of bulk solids for more than 15 years throughout the world. The main technology is the Vibrating Electrical Elevator and Reactor (REVE) which is based on three simple principles:

- The product is transported and mixed in a stainless steel spiral tube by vibrations induced by two off balanced motors.
- This stainless steel spiral pipe is connected to a power transformer and heated by electricity flowing in the wall of the pipe (impedance tube). The product is then heated up by contact with the hot tube.
- During transport in the confined spiral tube, the atmosphere is controlled: steam or specific gases can be injected on one side of the reactor and air/moisture/gases can be extracted on the other side.

This technology has been validated on several products such nuts, seeds, herbs, spices, flour or cereals. Figure 1 is an example of a REVE unit for food processing.

Figure 1: Example of food processing unit

Using the pilot unit of the Research and Development department, solid hydrodynamic characterizations were carried out, as well as food application measurements. Plug flow behavior of the REVE has been validated by temperature sensors moving inside the reactor and runs with colored wheat flour. It has been shown that the value of *D/uL* is ranging 10^{-3} to 10^{-5}, depending on product velocity, which assimilates the REVE to a plug flow reactor [2]. This characteristic ensures a homogeneous products treatment inside the REVE.

Pasteurization application has been validated on in-shell pistachios inoculated with *Enterococcus faecium NRRL B-2354* as a surrogate organism for *Salmonella* [3]. Tests were conducted with a pistachios flowrate of 100 kg/h with tube temperature ranging 110-120°C, steam flow ranging 5-10% and residence time ranging 5 to 11 minutes. With the increase of all those three parameters, there is an increase of the CFU log reduction up to more than 5.41, which is higher than the Almond Board of California (4-log) and even the FDA (5-log) recommendations [4].

Malt roasting tests, as an example of value-added food processing, were also conducted in the same pilot unit. Without steam injection in this case, 200 kg/h of malt have been processed at 240°C, 243°C, 245°C and 250°C with a residence time up to 80 minutes. For every temperature, there is an increase of EBC coloration with residence time until a critical value from which the EBC is decreasing drastically. During these tests, critical residence times were approximately 40 min, 35 min, 25 min and 20 min for respectively 240°C, 243°C, 245°C and 250°C. A dark malt (1700°EBC) was obtained after 40 min at 240°C, which is quicker than typical malt roasting processes (120 minutes) [5].

Thanks to the pilot unit, several food thermal processing can be tested and validated, revealing the versatility of the REVE in food industry. From trial results, continuous industrial units are designed and installed up to 5 ton/h. For future works, new applications/products have to be tested with for instance the use of specific gases.

[1] Fellows, P. (2000). Food processing technology, Principles and Practice. Woodhead Publishing Limited. 229-384.

[2] Levenspiel, O. (1972). Chemical reaction engineering. Wiley.

[3] Almond Board of California (2007). Guidelines for Process Validation Using Enterococcus faecium NRRL B-2354.

[4] Almond Board of California (2007). Outgoing Quality Control Rule. Federal Register, 72 (61): 15021-15036.

[5] Samaras, T., Camburn P., Chandra, S., Gordon, M., and Ames, J. (2005). Antioxidant Properties of Kilned and Roasted Malts. J. Agric. Food Chem, 53: 8068-8074.

349185 Upgrading a CDU Main Fractionator with Advanced Mass Transfer Components

Abstract:

Maximizing capacity while improving efficiency and reliability of major equipment, such as distillation columns, is a very challenging target.

This is what most of the European Refineries are looking for due to the low refinery margins they are facing with, and the high competitiveness of the big Refineries of the Middle East and India

This paper will describe in details the revamp of a Crude Distillation Main Fractionator of a major Refinery in West Europe, where the use of advanced mass transfer components i.e. High Performance Trays, High Performance Grids, and tailor made tower internals allowed to achieve challenging targets and operation reliability with minimum investment cost.

349942 Tomography Gamma-Scan Applications for Packed Towers

Abstract:

Packed columns are widely used in distillation, absorption and stripping processes. It is well known that liquid maldistribution can severely reduce column efficiency and operability. Maldistribution creates localized disparity between the desired liquid/vapor ratios; reduces the effective packing efficiency for mass and heat transfer, translating to a high HETP and causes under-irrigated or dry areas that can result in localized coking or sintered chemical deposits within the packed bed.

For packed towers a grid-scan, 3 or 4 equal-distant scans would be done to qualitatively measure the liquid distribution. While this gamma scan procedure is fine for screening to detect macro problems of severe liquid maldistribution there are more subtle patterns of maldistribution better diagnosed by a more comprehensive method.

A tomography scan provides cross-sectional images of three-dimensional (3D) objects. The most prevalent use of tomography is still in medical applications, for instance the Computerized Tomography (CT) device that is used for non-destructive examination of the human body. Gamma-ray tomography takes advantage of the penetrating properties of high-energy gamma radiation to generate a high-resolution cross sectional density profile through a packed bed.

This paper will describe gamma-ray computer-aided tomography (CAT-Scan) for measuring the density profile across a packed bed showing its liquid distribution. Applications of CAT-Scans for diagnosing, monitoring and solving maldistribution and coking problems in refinery Vacuum and FCC Main Fractionating towers will be included in the presentation.

351127 Using Distributed Feed to Assess Potential Commercial Gains Realisable in an Atmospheric Crude Distillation Unit

Abstract:

Abstract
Distillation is an energy intensive process which consumes an estimated 3% of the world's energy (Masoumi & Kadkhodaie, 2012). The ever increasing interest in environmental issues and dwindling energy resources has impelled engineers in the petrochemical and allied industries to synthesise innovative low-cost distillation designs that are energy efficient. One such model which has been proposed and shown to have great promise for non-sharp split separations is utilisation of multiple feeds (Beneke et al., 2012). The objective of this investigation was to assess the potential commercial gains with regard to energy savings and capital expenditure achievable through use of distributed feed in an Atmospheric Crude Distillation Unit (ACDU). Prior to applying distributed feed to an Aspen Plus simulated ACDU, the method was initially tested on a three component ideal mixture of pentane heptane and hexane under the supposition that it would behave in the same manner as the synthetic crude feed. Column Profile Maps (CPM) in conjunction with Aspen Plus was used to synthesise and check for the effectiveness of the technique. The same concept was then advanced to the ACDU taking in a synthetic crude feed in an Aspen Plus environment. The results showed that, there was major potential to reduce the overall energy consumption of the ACDU. There were energy savings realized in the condenser as well as the furnace duties of up to 30% and with it the potential to reduce the said equipment size, thus both operating expenses and capital expenditure could be minimized.
References
1. Beneke, D., Peters, M., Hildebrandt, D. & Glasser, D., 2013. Understanding Distillation Using Column Profile Maps. New York: John Wiley & Sons.

2. Masoumi, M. E., Kadkhodaie, S., 2012. Optimization of Energy Consumption in Sequential Distillation Column. World Academy of Science, Engineering and Technology, 61, pp.576-80

351639 Characterization of Spray Nozzle Performance and De-Entrainment for Industrial Spray Applications

Abstract:

Spray nozzles are commonly implemented in refining, chemical, and process engineering applications, yet minimal fundamental knowledge exists for these systems. Specific applications for petroleum refining include crude and vacuum towers where predicting droplet size and entrainment is crucial for effective design and operation. Furthermore, the availability of experimental data is typically limited to air-water systems. For these applications, both spray performance and entrainment are key issues in the proper design of these systems, especially in the presence of countercurrent gas flow. This work utilizes Phase Doppler Interferometry (PDI) to characterize entrainment effects in a variety of systems including air-water and air-amine systems. The PDI system utilized in this work has the ability to interrogate concentrated dispersions and provide fundamental insight about entrainment. The PDI is coupled with an 8 in. glass column with the ability to process a variety of fluids at high gas rates (C-factors ~0.5 ft/s). A comprehensive analysis of entrainment is presented that focuses on nozzle selection, gas flow rate, and the effectiveness of de-entrainment devices. A variety of spray nozzle orientations/positions is evaluated with and without de-entrainment devices.

DISTILLATION FUNDAMENTALS

By

Mike Resetarits

Anand Vennavelli

AY Ogundeji

Fractionation Research Inc.
424 S. Squires Street, Suite 200
Stillwater, OK 74074

2 April 2014

Presented at YP Tutorials: Distillation and Absorption Equipment
AIChE Spring Meeting, New Orleans, Louisiana

Abstract

In the chemicals processing industry, distillation is the most commonly employed separation technique. Different chemicals have different boiling points. After a simple mixture flash, the vapor product contains more light compounds than does the liquid product. A series of flash vessels can be employed to effect purifications of the light and heavy compounds. Similarly, a series of equilibrium stages arranged vertically in a distillation column can effect simple and difficult separations. Column operating pressure is an important variable as is column pressure drop. A reboiler is required at the bottom of a column to boil the heavy liquid. A condenser is required at the top of a column to condense the light vapor and create a reflux stream that flows down the column and cleans the upflowing vapor. When the reflux flow rate is divided by the top (liquid) product rate, the resultant number is called the reflux ratio. Generally, reflux requirements are a function of the relative volatilities of the light and heavy components. Pumps, pipes, valves, meters, thermocouples and other instrumentation are all essential parts of a distillation unit. Batch, azeotropic and extractive distillations are also employed industrially.

Keywords: Distillation Separation Boiling Flashing

Acknowledgments

This tutorial paper is dedicated to the thousands of global engineers and technicians who design, build, operate and maintain distillation columns.

Introduction

When a liquid mixture is fully boiled, to dryness, the resultant vapor will have the exact same composition as the starting liquid. However, when a liquid mixture is only partially boiled, the lighter compounds (usually those with lower molecular weights) will leave the liquid first. The resultant liquid will be richer in heavier compounds (usually those with higher molecular weights). A partial separation has occurred, but the resultant vapor will not be pure and the resultant liquid will not be pure.

Now imagine a mixing chamber or a vertical column wherein a light liquid stream encounters a heavy vapor stream. Also imagine that the streams are not at compositional equilibrium. Some of the light components in the liquid stream might leave it and enter the vapor stream. Similarly, some of the heavy components in the vapor stream might leave it and enter the liquid stream. If, as the streams encounter each other, compositional equilibrium is achieved, it is said that an "equilibrium stage" or a "theoretical tray" of contacting has occurred. Such a compositional equilibrium will be achieved if the liquid and vapor phases contact each other intimately for a sufficient amount of time. When the equilibrium is attained, further contacting or mixing does not yield additional mass transfer.

Distillation begins with the boiling of a mixture. A distillation column can be viewed as a series of equilibrium stages arranged vertically. A liquid stream flows down from stage to stage and encounters a vapor stream that flows up from stage to stage. As compositional equilibria are attained in each of those stages, the downward-flowing liquid stream becomes heavier and more pure as the upward-flowing vapor stream becomes lighter and more pure. Distillation columns are capable of starting with a mixture and turning it into a very pure top product rich in light compounds, and a very pure bottom product rich in heavy compounds.

Discussion

Boiling Points

The science of distillation begins with the study of boiling points. The temperature at which a liquid boils depends upon the pressure that the top liquid surface is subjected to. When that pressure is atmospheric pressure, the boiling point is referred to as the "normal boiling point." **Table 1** gives the normal boiling points of several compounds of the "aromatics family" of chemicals. All of these compounds exist as liquids at room temperature and pressure. For distillation separations to occur, mixtures of these compounds need to be partially boiled. **Table 2** includes the normal boiling points of the main compounds that are found in a natural gas processing plant. All of these compounds exist as vapors at room temperature and pressure. For distillation separations to occur, mixtures of these compounds need to be partially condensed. Distillation always requires the presence of two fluids, a liquid and a vapor. In both of these tables, higher molecular weight compounds have higher boiling temperatures.

Bubble and Dew Points

The separation of benzene (B) from toluene (T) is an important industrial separation. It is usually effected via distillation. At atmospheric pressure, pure benzene boils at 80°C; pure toluene, at 111°C. A mixture of benzene and toluen e begins to boil across a wide range of temperatures; depending on the mixture composition. The lower curve of **Figure 1** gives the bubble point temperatures. Very unfortunately, taking any mixture of benzene and toluene, and raising its temperature to 80°C or higher will not simply boil off all of the benzene, leaving pure toluene behind. When boiling most mixtures, the lighter component indeed boils first, but the resultant vapor includes very significant quantities of the heavier component.

Referring back to **Figure 1**, regarding B and T, the lower curve represents mixture bubble points. The upper curve represents mixture dew points. The area between the curve is the so-called "2-phase region."

Assume a cold 50/50 BT mixture (Point m_L in **Figure 1**) and assume that its temperature is being raised continuously. It will not begin to boil at 80°C; instead the bubble point will prove to be 92°C (Point b_1 in **Figure 1**). The composition of the first bubble of vapor is represented by Point d_1. It is no surprise that d_1 is high in B content. As the 50/50 2-phase mixture is heated further, when the temperature reaches 95.5°C, rough ly half of the mixture will be vaporized. The vapor composition is represented by d_2; the liquid, b_2. The horizontal lines in **Figure 2** are called "tie lines." At this point, some separation of the two compounds has indeed been achieved, but the two phases are far from being pure. With additional heating of the mixture, the dew point (d_3) will be achieved at a temperature of 99°C. The v ery last droplet of liquid will have the composition represented by b_3.

If the mixture is heated all the way to 110°C, the resultant vapor (Point m_V on the graph) will have the same compositions as the initial liquid (m_L).

As the BT mixture was being heated, regardless of the temperature, there never came a point where the B or the T approached "purity" – not in the liquid phase and not in the vapor phase.

Flash Distillation

Figure 2 summarizes a simple flash distillation, again using the example of a 50/50 B/T liquid mixture. Imagine that the liquid mixture is heated to a temperature of 95.5°C with a subsequent let-down to atmospheric pressure. **Figure 2** shows that a modest amount of separation has occurred. In fact, the after-flash vapor and liquid compositions are represented by Points b_2 and d_2 in **Figure 1.**

Figure 3 shows a series of flash distillations, where the resultant liquids are flashed again and again at ever-increasing temperatures. After 5 flashes, a T purity of 91% has been achieved. (For an industrial benzene column, the T purity target is closer to 99% than 90%.) Most importantly, **Figure 3** shows that a series of operations, in this case flashing, can be used to separate and purify chemical compounds that have different boiling points.

In **Figures 1**, **2** and **3**, boiling was always effected at atmospheric pressures, with boiling temperatures in the neighborhood of 80 to 110℃. L ow pressure steam at 2.5 bara (36 psia) and 127℃ could be employed in the boilers (heat ex changers).

Vapor Pressures and Relative Volatilities

Generally, with a pure liquid, molecules are constantly escaping the top liquid surface and are becoming vapor. The pressure that those evaporated molecules exert is called the "vapor pressure" of the pure liquid. When a pure liquid is heated, the vapor pressure of the liquid increases. When the vapor pressure of the pure liquid equals atmospheric pressure, the "normal boiling point temperature" is achieved and evaporation becomes violent, i.e., the pure liquid "boils."

Different liquid compounds have different vapor pressures, and thus, different normal boiling points. Generally, smaller molecules have higher vapor pressures. When a liquid-vapor mixture is enclosed and heated, as the pressure of the vessel rises, the boiling point temperature also rises. If a vacuum is created above that mixture, the boiling temperature of the mixture becomes less than the normal (atmospheric pressure) boiling temperature. As an example, water can boil across a very broad range of temperatures. A very simple laboratory experiment can show how easily water can boil at room temperature. **Table 3** gives details regarding how pressure affects the boiling temperature of water.

Consider a mixture of small(1) and large(2) molecules. When that mixture is heated, and boiled, the smaller molecules will usually have higher vapor pressures and will escape the liquid phase first. The vapor phase composition will have more lighter compounds than the leftover liquid phase; a partial distillation separation will have occurred.

The ratio of the vapor pressures (VP) of different chemical compounds is often called the "relative volatility" and is often represented by "∝". The following simple equations apply:

$$\alpha_{12} = \frac{VP_1}{VP_2}$$

$$y_1 = \frac{\alpha_{12} x_1}{1 + (\alpha_{12} - 1) x_1}$$

$$x_1 + x_2 = 1$$

$$y_1 + y_2 = 1$$

where

x_1 = composition of Compound 1 in liquid phase

x_2 = composition of Compound 2 in liquid phase

y_1 = composition of Compound 1 in vapor phase

y_2 = composition of Compound 2 in vapor phase

For many chemical compounds, their vapor pressures are known across broad ranges of temperatures. These vapor pressures can be employed within the above equations to determine the separations that are achievable by boiling mixtures of compounds.

Figure 4 provides the vapor pressures of benzene and toluene across a broad range of temperatures. Notice how the graphed data converge at higher temperatures. At higher temperatures, the relative volatilities of binary compounds do tend to decrease, i.e., the separations become more difficult. **Figure 5** provides the vapor pressures of propylene and propane.

Figure 6 illustrates the ramifications of relative volatilities (essentially, vapor pressure ratios) on vapor-liquid equilibria. Notice that at α =1.1, the composition of component 1 in the vapor and liquid phases are just about equal. At α =3.0, the composition of the light component in the vapor phase is much higher than it is in the liquid phase.

Column Configuration and Ancillary Equipment

Figure 7 is a diagram of a typical distillation unit including the column and the heat exchangers. Column diameters can be as small as 2 inches or as large as 50 feet. Column heights can range from 12 inches to 350 feet. The inside of the column can be equipped with trays or packings – to mix the countercurrent liquid and vapor streams.

Figure 7 shows a feed stream entering the column about half-way up; this is not unusual. The feed stream can be a liquid, a flashing liquid or a vapor. For the sake of this discussion, it will be assumed that the feed stream is a liquid. The feed enters the column and joins the down-flowing reflux liquid stream. As the combined liquid stream flows downward, mass is exchanged with an upward flowing vapor. At the bottom of the column, most of the liquid stream is boiled in a "reboiler". The resultant vapor stream is called the "boil-up." In **Figure 7**, steam is being used to boil the liquid. A portion of the bottom liquid is not reboiled and is the final "bottoms product." All of the fluids that reach the bottom of the column are rich in heavy components, i.e., chemicals with high boiling points. The vapor leaving the reboiler is similarly heavy. As this heavy vapor flows upwards through the column, it encounters a lighter liquid. There is a compositional disequilibrium between the phases which causes light components to leave the liquid and heavy components to leave the vapor. In fact, as the vapor stream rises past the feed point, it becomes richer in lighter components. By the time that the vapor reaches the top of the column it is quite light. In **Figure 7**, the entire vapor stream enters a condenser which totally converts the vapor stream to a liquid. Some of this liquid is returned to the column as a "reflux" and some becomes the "overhead product." All of the fluids near the top of the column are quite light, including the liquid reflux stream. As this light reflux stream flows down the column it encounters a vapor stream that is relatively heavier. A compositional disequilibrium exists and heavy components in the vapor stream leave the vapor stream. Light components in the liquid stream leave the liquid stream. Some people regard the reflux stream as "clean'; they regard the reflux stream as cleaning the upflowing vapor stream.

In **Figure 7**, cooling water is employed as the heat sink medium in the condenser. Air (in a fin fan condenser) or refrigerants can also be used. Condensers are not actually positioned at the tops of columns. They are positioned about 30 feet above ground level. In fact, reflux drums are typically positioned underneath most condensers to collect condensates. Pumps are required to move the reflux liquids back to the tops of the columns. **Figure 7** depicts a typical distillation unit. There are many variations.

Reflux Requirements

"Reflux ratio" is defined as the reflux flow rate divided by the overhead product rate. Difficult separations, with low relative volatilities, require higher reflux ratios. Higher reflux ratios require more boil-up in the reboilers and higher steam rates and higher cooling water rates in the condensers and, overall, higher energy consumption rates. Additionally, higher reflux ratios lead to increased vapor and liquid traffics inside the column and the column diameter must be increased when the column is being designed. Using the benzene-toluene binary as an example, **Figure 8** shows that taller columns, with more theoretical stages, can lead to reduced reflux ratios. **Figure 8** also shows that as top and bottom purity targets are increased, so are the needs for additional column height and refluxing.

Atmospheric Distillation

Whenever possible, distillation columns are designed for, and operated at, near-atmospheric pressure. **Table 1** provided the atmospheric pressure boiling points of several compounds of the "aromatics family." When these particular chemicals are distilled, low-grade low-temperature steam can be employed in the reboilers. At the same time, cooling water or air can be used in the condensers to effect liquefactions. There are other occasions, and chemicals, where the associated distillation columns need to be run at high pressures or at vacuum conditions. Such columns are more expensive to build and operate. High pressure columns require many extra safety considerations.

Table 1 shows that the boiling points of benzene (B) and toluene (T) are quite different. Indeed, these compounds can be separated and purified via boiling – and distillation. Benzene columns are quite common, globally, and are operated at near-atmospheric pressures. **Table 1** includes other aromatic compounds which will be referred to later.

High Pressure Distillations

Distillation separations cannot occur with chemical compounds that are totally in the liquid state, or totally in the vapor state. Liquid chemicals need to be partially vaporized in reboilers. This is usually "easy" and can be effected using steam, hot water, etc. Vapor chemicals need to be partially or totally condensed in condensers and this often creates "difficulties." Cooling water and air are the most common heat sinks for hot chemicals. Cooling water and air are commonly assumed to be at 100℉ = 38℃. The chemical compou nd vapors that they condense must therefore be at a higher temperature, like 130℉. As an alternative, a refrigerant could be employed as the heat sink fluid, but refrigeration is generally expensive and is generally avoided.

See **Table 2**, and in particular, the boiling points at atmospheric pressure, of pure methane and pure ethane. A natural gas plant demethanizer separates methane from ethane using distillation. If such a demethanizer is operated at atmospheric pressure, the methane-ethane mixture would need to be cooled to temperatures lower than -100℃ to create a liquid phase. Instead, demethanizers are typically operated at pressures like 500 psia, to facilitate liquefaction. At pressures like 500 psia, the bubble and dew points of methane-ethane are high enough that cooling water and air can be used in the distillation column condenser.

Referring again to **Table 2**, this time noting the (atmospheric pressure) boiling temperatures of propylene and propane. The separation of propylene from propane is an important global process for polypropylene production. If a propylene-propane distillation column were run at atmospheric pressure, a (costly) refrigerant at about -50℃ would be required as the heat sink in the overhead condenser.

Figure 5 provides vapor pressure data for these C_3 compounds. Most propylene-propane columns run at pressures like 300 psia. At such pressures, per **Figure 5**, these chemicals condense at temperatures like 130℉. Cooling water or air (at 100℉) can be used to effect these liquefactions in overhead condensers.

Vacuum Distillation

The polymer known as polystyrene is created using relative pure styrene monomer. Plants that produce styrene include distillation columns that separate ethylbenzene (EB) from styrene (S). Refer back to **Table 1** and note the atmospheric-pressure boiling points of pure ethylbenzene and pure styrene, 136 and 146℃, respectively. If an EB/S column were operated at atmospheric pressure, those chemicals would be subjected to such temperatures. Very unfortunately, such temperatures caused styrene monomers to combine repeatedly to form polymer molecules and those molecules solidify. Such atmospheric-pressure EB/S columns would plug up (foul) quickly. Instead, EB/S columns are actually operated at very deep vacuums, like 100 mmHga, to reduce the column temperatures and virtually eliminate the renegade polymerizations and the plugging.

There are other reasons why some distillation columns are operated at sub-atmospheric pressures and relatively low temperatures. At atmospheric pressures and higher temperatures, some molecules crack apart instead of boil. As an example, the second major distillation column in most refineries is typically run at vacuum pressures (and lower temperatures) to prevent large oil molecules from cracking.

Unpractical Distillations

Some distillation separations are quite difficult, but they are performed industrially nevertheless. Refer again to **Table 1**. The boiling points of EB and p-xylene are quite close, and yet, this distillation is performed industrially, but rarely. This is also true of the distillation of p- and m-xylene from o-xylene. Conversely, however, the boiling points of p-xylene and m-xylene are so close (0.8℃, actually) that this distillation is n ever performed industrially. As it turns out, the melting points of these two compounds are amazingly different (+13 and -47℃, respectively) and crystallization can be employed to separate them.

Closing

Thirty years ago, Dr. James Fair of the University of Texas at Austin, predicted that distillation would be a popular separation method for many decades. He was correct!

Figure 1: Benzene - Toluene Bubble and Dew Points; Pressure = 14.7 psia

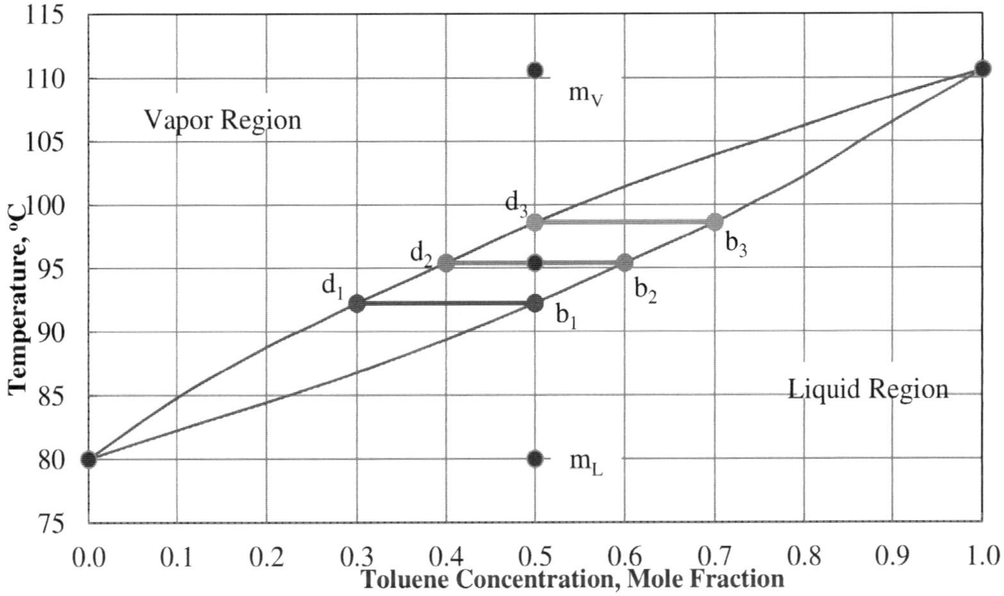

Figure 2: Flash Distillation

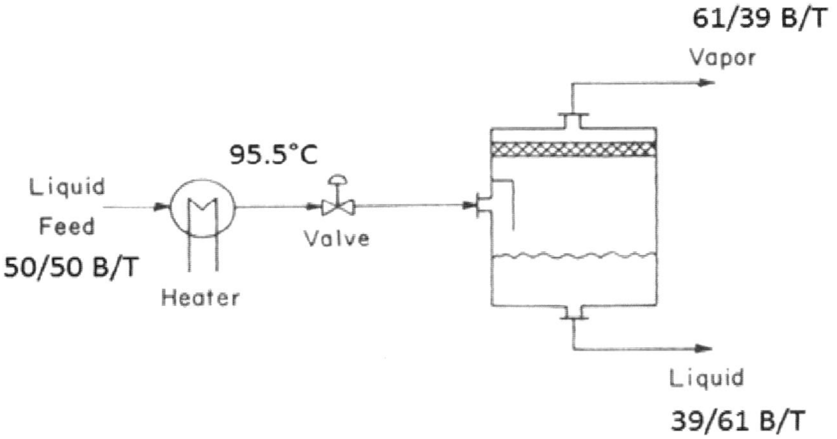

Figure 3: Successive Flash Distillations

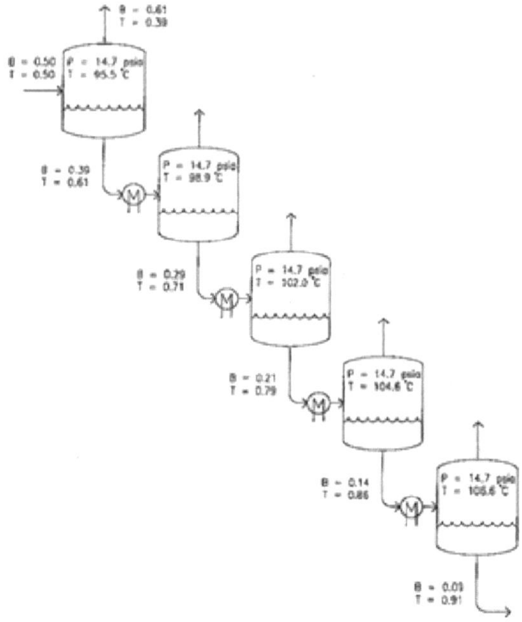

Figure 4: Vapor Pressures of Benzene and Toluene

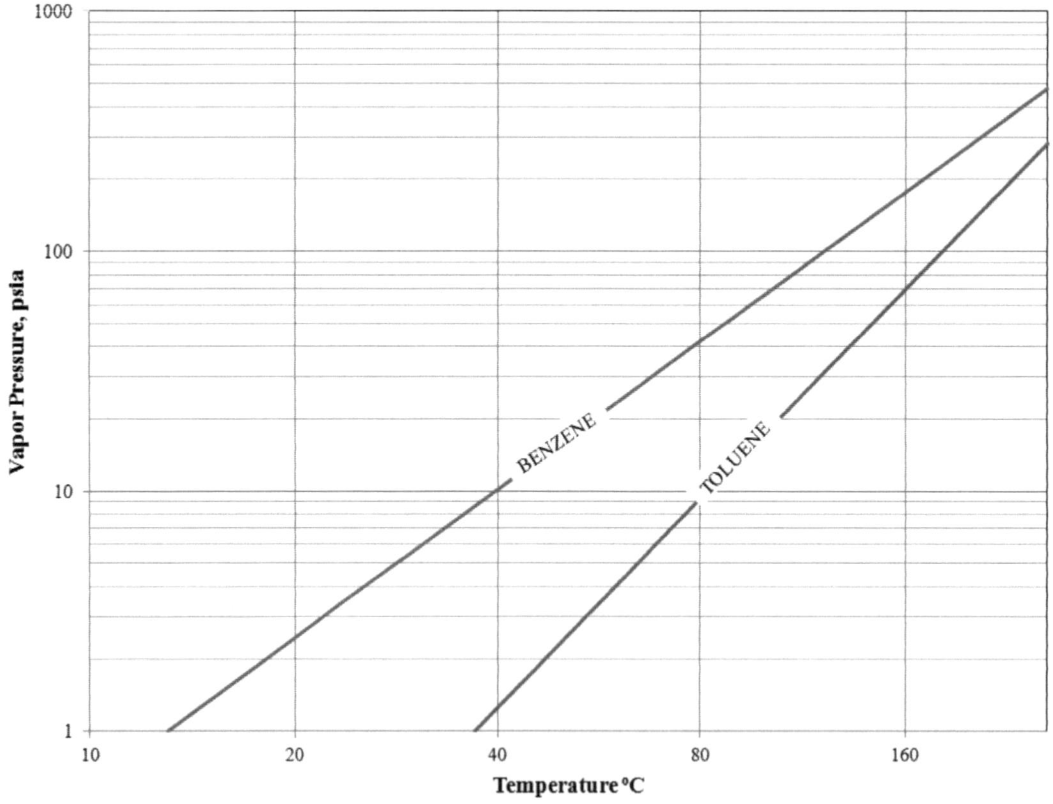

Figure 5

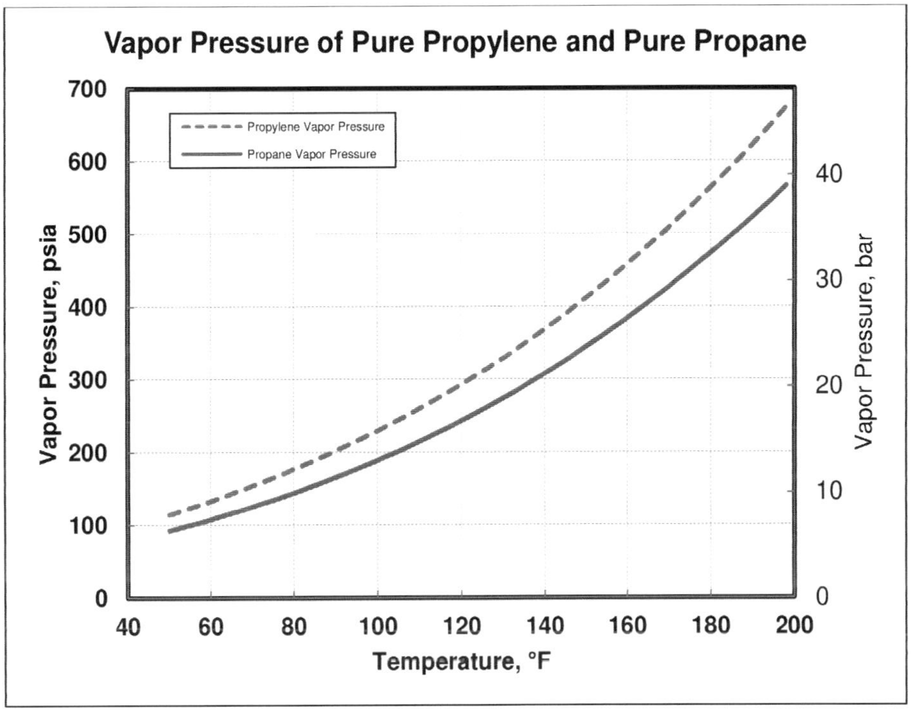

FIGURE 6 Vapor (y_1) versus liquid (x_1) concentration as a function of relative volatility.

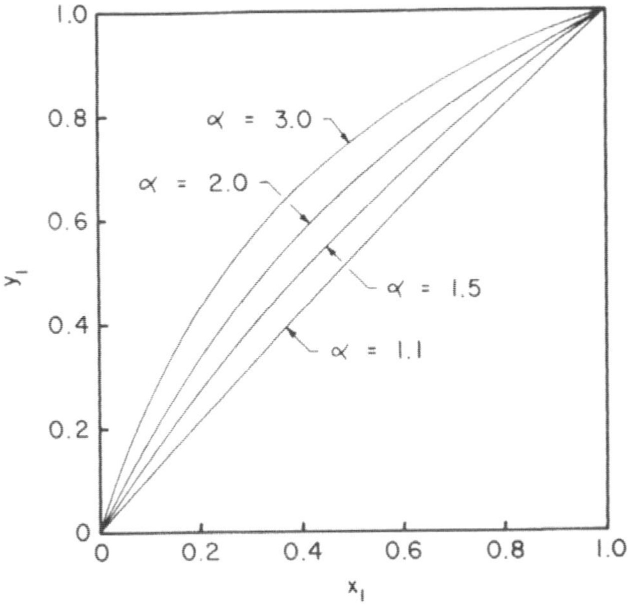

FIGURE 7 Schematic representation of a distillation column.

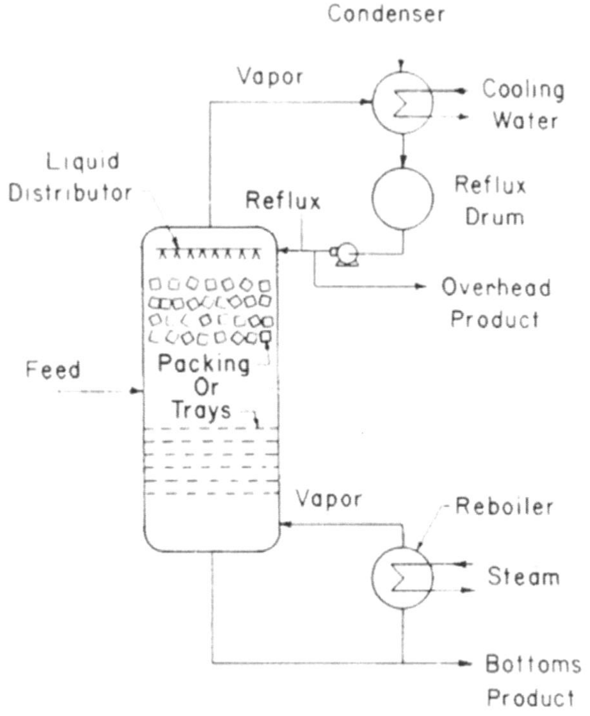

FIGURE 8 Theoretical stages versus reflux ratio (benzene-toluene at 1.0 bar). x_D, mole fraction benzene in overhead; x_B, mole fraction toluene in bottoms.

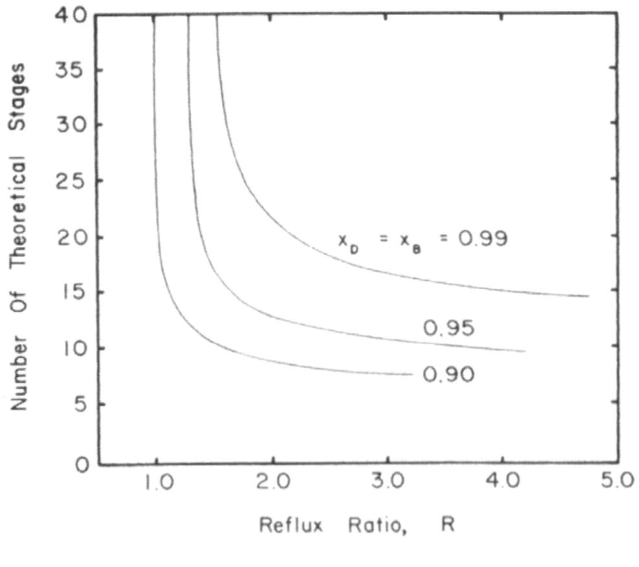

Table 1
Boiling Temperatures at Atmospheric Pressure

	Boiling Point, °C
Benzene (B)	80
Toluene (T)	111
Ethylbenzene (EB)	136
p-Xylene	138
m-Xylene	139
o-Xylene	144
Isopropylbenzene	153
Styrene (S)	146

Table 2
Boiling Temperatures at Atmospheric Pressure

	Boiling Point, °C
Methane	-161
Ethylene	-104
Ethane	-89
Propylene	-48
Propane	-42
Isobutane	-12
n-Butane	-1
n-Pentane	+36

Table 3
Water Boiling Temperatures

Pressure, psia	Temp, °C	Temp, °F
1.0	39	102
14.7	100	212
100	164	328
600	252	486

REFERENCES

1. Kister, H.Z., Distillation Design, McGraw-Hill, NY, NY, 1992.
2. Kister, H.Z., Distillation Operation, McGraw-Hill, NY, NY, 1990.
3. Lockett, M.J., Distillation Tray Fundamentals, Cambridge Univ. Press, NY, NY, 1986.
4. Strigle, R.F., Packed Tower Design and Applications – Random and Structured Packing, 2nd Edition, Gulf Publishing Co., Houston, TX, 1994.
5. Resetarits, M.R., Agnello, J., Lockett, M.J., and Kirkpatrick, H.L., "Retraying Increases C3 Splitter Column Capacity," Oil and Gas Journal (86), June 1988, pp. 54.
6. Resetarits, M.R. and Miller, R.J., "Increase Demethanizer Efficiency and Capacity," Hydrocarbon Processing, October 1991, pp. 79-81.
7. Resetarits, M.R., Navarre, J.L., Monkelbaan, D.R., Hangx, G.W.A. and van den Akker, R.M.A., "Trays Inhibit Foaming," Hydrocarbon Processing, March 1992, pp. 61-64.
8. Resetarits, M.R., Riter, J.B. and Weiland, R.H., "Distillation Technology: Cocurrent Flow Trays in Distillation Columns," Hydrocarbon Engineering, Vol. 7, No. 9, September 2002, pp. 85-88.
9. O'Connell, H.E., Plate Efficiency of Fractionating Columns and Absorbers, Trans. Am. Inst. Chem. Engrs., 42, 741, 1946.
10. Resetarits, M.R., Miller, R.J., Navarre, J.L., Linskeseder, M. and Reich-Rohrwig, P., "New Enhanced Capacity MD Tray Debottlenecks Deethanizer," IChemE Distillation and Absorption Conference, Birmingham, U.K., September 7, 1992.
11. Resetarits, M.R., Monkelbaan, D.R. and Miller, R.J., "Trays in Rocking Services," IChemE Distillation and Absorption Conference, Birmingham, U.K., September 7, 1992.
12. Resetarits, M.R., Urbanski, N.F., and Xu, Z.P., "Column Revamps," AIChE Spring Meeting, Chicago, IL, February 16, 1998.
13. Resetarits, M.R. and Pappademos, N., "Factors Influencing Vapor Crossflow Channeling," AIChE Annual Meeting, Reno, NV, November 8, 2001.
14. Resetarits, M.R. and Lockett, M.J., "Distillation," Encyclopedia of Physical Science and Technology, Vol. 4, 2001.
15. Resetarits, M.R. and Ogundeji, A.Y., "On Distillation Tray Weir Loadings," AIChE Spring Meeting, Tampa, FL, April 27-29, 2009.
16. Resetarits, M.R., Cai, T.J. and Whiteley, J.R., "On Distillation Tray Entrainment", AIChE Spring Meeting, Chicago, IL, March 2011.
17. Chambers, S., Cai, T.J., Vennavelli, A.N., Ogundeji, A.Y., Resetarits, M.R. and Redden, J.C., "Effect of Tray Spacing on Tray Performance," AIChE Spring Meeting, April 3, 2012.

Trays for Distillation and Absorption

By

AY Ogundeji
Mike Resetarits
Tony Cai

Fractionation Research Inc.
424 S. Squires Street, Suite 200
Stillwater, OK 74074

2 April 2014

Presented at YP Tutorials: Distillation and Absorption Equipment, AIChE Spring Meeting, New Orleans, Louisiana, April 2, 2014

Abstract

Distillation columns are essential to the chemical processing industries. They need to be properly designed and operated. Column internals, such as trays and packings, are essential to column performance. If column internals malfunction, the entire plant may suffer. Among column internal, trayed columns are widely used in distillation and absorption applications. This presentation will discuss the ranges of the liquid and vapor loads that a trayed column can handle. It will review the systems where trayed columns are particularly applicable. A brief history of trays and trayed columns will be presented. New generations of high performance trays, including augmented crossflow, counter current flow and cocurrent flow trays, will be discussed. The performance differences between trays and packings will be summarized. Column designs that provide high energy efficiencies will be briefly discussed. Extensive video footage of tray operations will be shown, from commercial size columns with various distillation systems ranging from deep vacuum to high pressure.

Keywords: Distillation Absorption Tray Flooding

OVERVIEW

Distillation is a process of separating mixtures into two or more products based on differences in their boiling points in a boiling mixture. The distillation process has been around for hundreds of years and, it is one of the most widely used processes in the industries.

Absorption is used to separate gas mixtures, remove impurities, or recover valuable chemicals. In absorption, there is a transfer of one or more species from the gas phase to a liquid absorbent. The species transferred to the liquid absorbent are referred to as solutes or absorbates.

Distillation and absorption employ trays and packings as column internals. This paper focuses only on tray internals. These column internals are used as phase contacting devices. There are different types of trays. The table below shows a rough comparison of the common tray types.

Type	**Sieve Tray**	**Valve Tray**	**Bubble-Cap Tray**	**Dual-Flow Tray**
Capacity	High	High	Moderate	Very high
Efficiency	High	High	Moderate	Lower than others
Pressure drop	Moderate	Moderate	High	Low to moderate
Turndown	About 2:1	About 3:1	Highest	Low
Cost	Low	About 20% higher than sieve	High. About 2-3 times more than sieve	Low

GENERAL

For distillation, a liquid phase flows countercurrently with a vapor phase, while for absorption, a liquid phase flows countercurrently with a gas phase. This is shown in **Figure 1**.

Figure 2 shows a drawing of a simple distillation column with trays. The figure shows that the column contains 20 trays. Feed and reflux streams are introduced into Tray 10 and Tray 20, respectively. Vapor is introduced into the bottom of the column via the "From Reboiler" nozzle.

The vapor rises and mixes with the liquid flowing down the column. The liquid flowing downward exits the column via the "To Reboiler" piping while the upward flowing vapor exits the column via the "To Condenser" nozzle.

Figure 3 shows the different tray configurations. Multi-pass trays are usually employed when liquid rates are too large for a 1-pass tray.

GENERATION OF TRAYS

There are different generations of trays. The generations include crossflow, augmented crossflow, countercurrent and cocurrent flow trays. The first generation is simple crossflow trays which include bubble-cap, sieve and valve trays, and which have the least capacities. The cocurrent flow trays (Shell Consep, Koch-Glitsch UltraFrac and UOP SimulFlow trays), which are the latest generation and have the highest capacities.

TRAY PARTS

Figure 4 shows a drawing of a 1-pass crossflow tray. It shows different parts of a tray. Liquid flows across each tray deck, over an outlet weir, and into a downcomer. The outlet weir purpose is to build a liquid head on the tray deck. A receiving pan collects and redirects the liquid onto the next deck through the downcomer clearance. The vertical distance between successive trays is called "spacing". The inner side of the downcomer is a vertical metal plate called the curtain. The column wall is also known as the shell.

Figure 5 shows a simplification of 1-pass crossflow tray hydraulics. Liquid enters the tray deck from the downcomer of the tray above by gravity. The liquid on the tray is aerated with vapor rising through the deck perforations to form a froth. The froth flows across the tray deck, over the outlet weir and into the downcomer where the vapor disengages. The distance between the top of the outlet weir and the top surface of the liquid flowing over the weir is called crest.

Figures 4 & 6 can be used to compare conventional and augmented crossflow trays, respectively. Augmented crossflow trays have truncated downcomers. That is, the downcomers have bottoms, and the bottoms are perforated. Augmented crossflow trays also have perforated decks below the bottoms of the downcomers as shown in Figure 6. Conventional crossflow trays do not have truncated downcomers. The higher open area of augmented crossflow trays yields higher capacities than conventional crossflow trays.

Figure 7 shows different types of tray decks. The choice of deck usually depends on the application. Vapor-liquid handling capacities and pressure drops of decks differ. For example, bubble-cap decks turn down better than sieve and valve decks. Bubble cap trays have higher pressure drops than sieve and valve trays. Moving valve decks have larger rangeabilities than simple sieve decks. Push valves reduce froth heights and increase capacities.

FLOODING

The performance of trayed columns is determined by many factors that include flooding, weeping and dumping. Flooding is excessive accumulation of liquid inside a trayed tower. When trays flood, there is sudden increase in column differential pressure and decrease in mass transfer efficiency. There are many different mechanisms that cause trays to flood. These mechanisms include choking, downcomer backup, spray entrainment and froth height flooding. More details about the mechanisms are as follows:

Choke Flooding

It occurs when the liquid flow rate increases to a point where the velocity of the biphase in the downcomer exceeds a certain limit. At this point, the biphase cannot even fit into the downcomer mouth and the biphase cannot be transported to the tray below. It happens when the downcomer top area (mouth) is too small to accommodate the aerated liquid above the outlet weir. A sketch is provided in **Figure 8**.

Downcomer Backup Flooding

It occurs because of high tray pressure drop, high liquid height on the tray and large downcomer clearance losses. When the backup of biphase in the downcomer exceeds the tray spacing (plus the outlet weir height), liquid accumulates on the tray above.

Spray Entrainment Flooding

It happens at low liquid rates when vapor velocity is increased to a point where the liquid on the tray is dispersed by the rising vapor into liquid droplets which are entrained to the tray above.

Froth Height Flooding

It occurs at high liquid and vapor rates. If the tray spacing is 24 inches or less, as vapor rate increases, a point is reached where the froth height equals the tray spacing. Liquid accumulates on the tray above.

Weeping

This phenomenon occurs at low vapor rates. At low vapor rates, when the pressure exerted by the vapor is insufficient to hold up the liquid on the deck, liquid starts to leak through the deck perforations. This liquid backmixing yields reduced mass transfer efficiency.

Dumping

It occurs at very low vapor rates. Excessive weeping will lead to dumping. With dumping, all the liquid on the deck will weep through the perforations. In this condition, no liquid flows over the outlet weir.

CLOSING

Roughly half of old and new columns contain trays; the other half, packing. Fundamental research regarding these contacting devices has been performed at Fractionation Research Inc. (FRI) since 1954.

REFERENCES

- H.Z. Kister, Distillation Design, McGraw-Hill, 1992.
- J.D. Seader, E.J. Henley and D.K. Roper, Separation Process Principles, John Wiley & Sons, 2011.
- M.J. Lockett, Distillation Tray Fundamentals, Cambridge University Press, N.Y., N.Y. 1986.

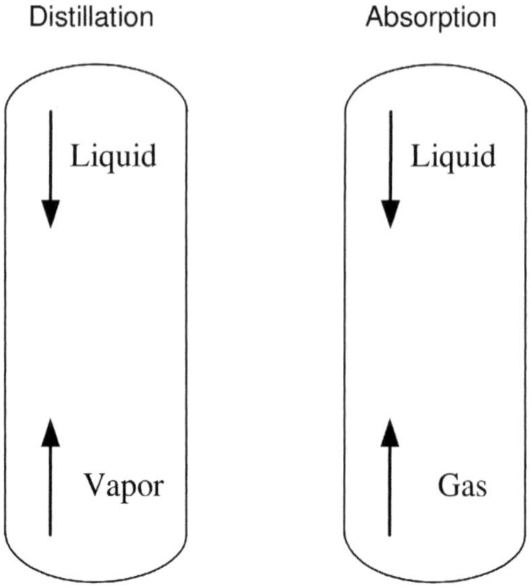

FIGURE 1.

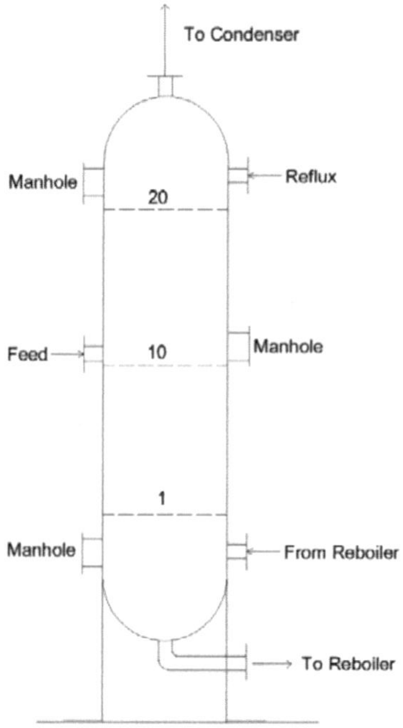

FIGURE 2. Simple Distillation Column with Trays

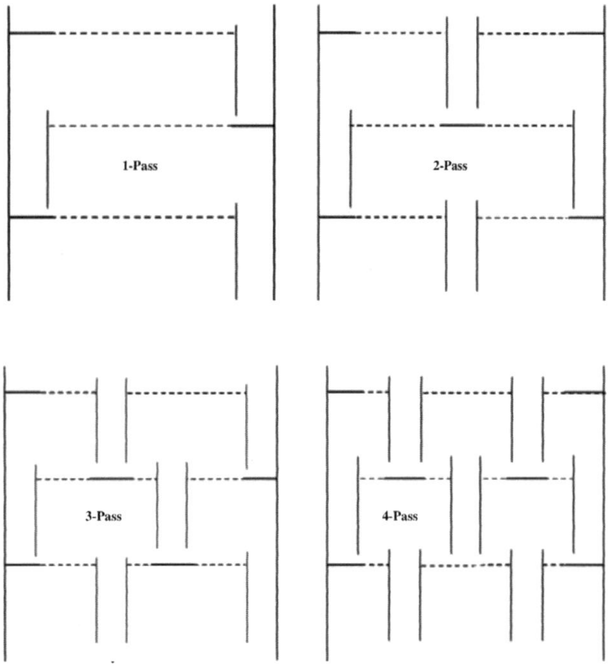

FIGURE 3. Crossflow Trays

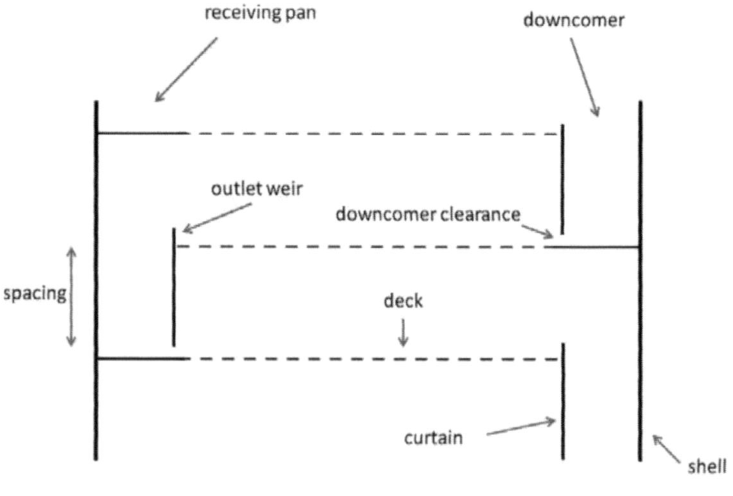

FIGURE 4. 1-Pass Crossflow Trays

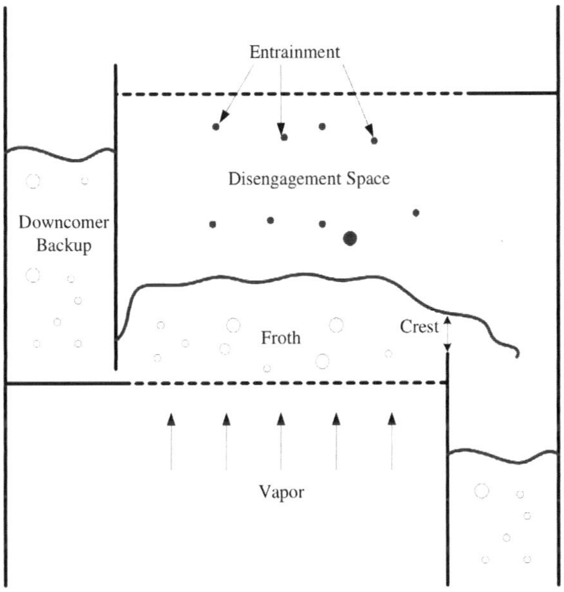

FIGURE 5. Flow Pattern on 1-Pass Crossflow Tray

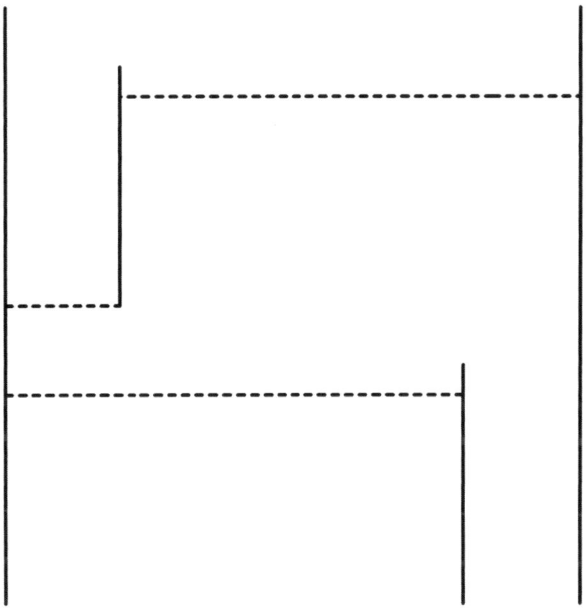

FIGURE 6. Augmented Crossflow Trays

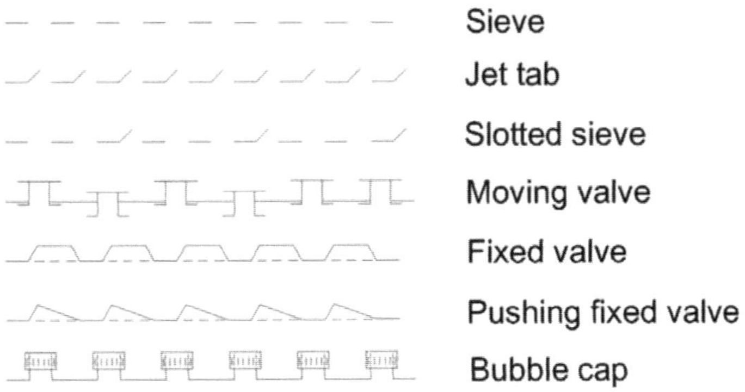

FIGURE 7. Tray Deck Types

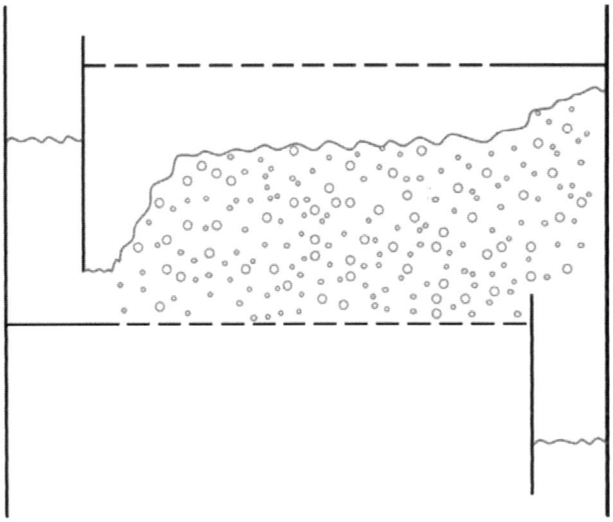

FIGURE 8. Choke Flooding

Structured Packings for Distillation and Absorption

By

Simon Chambers
Anand Vennavelli

Fractionation Research Inc.
424 S. Squires Street, Suite 200
Stillwater, OK 74074

2 April 2014

Presented at YP Tutorials: Distillation and Absorption Equipment,
AIChE Spring Meeting, New Orleans, Louisiana, April, 2014

UNPUBLISHED

Abstract

Structured packings used in industrial practice can be broadly classified into three generations: wire gauze, standard, and high capacity corrugated sheet metal. Standard structured packings were a popular choice in the industry for many years. Newer high capacity structured packings, which comprise either "half-S" or "full-S" crimp angle shapes, can operate at much higher load points or maximum useful capacities (MUC's), and considerably lower pressure drops than standard structured packings.

Structured packings are favored in high vapor rate, low liquid rate distillation services and in absorption. They are particularly advantageous in vacuum columns, where low pressure drops are sometimes critical. For example, in Ethylbenzene-Styrene monomer columns, low pressure drops yield low liquid temperatures and, consequently, reduced renegade polymerization.

In this paper, structured packing physical characteristics, installation, and applications are discussed. In addition, basic performance characteristics, such as HETP, pressure drop, and capacity, and liquid distribution quality are reviewed. A 20-minute video was prepared by FRI to illustrate the basics of structured packings.

Keywords: Packings Distillation Absorption Low Pressure Drop

1 Overview

Structured packings are often the choice of internals for low-pressure distillation and absorption. Structured packings provide good efficiencies, low liquid hold-ups, and low pressure drops.

Structured packings are frequently used in:

- Chemicals and fine chemicals (e.g., Ethylbenzene/styrene distillation)
- Refining (e.g., vacuum and atmospheric crude oil distillation)
- Petrochemicals

Structured packings are best suited for vacuum to moderate pressure applications.

2 Generations of Structured Packings

Structured packings can be broadly classified into three generations – wire gauze, standard corrugated sheet metal, and high capacity corrugated sheet metal.

2.1 First Generation: Wire Gauze Structured Packings

First introduced in the late 1950's to early 1960's, these packings are most suitable for difficult separations and when there is a low pressure drop requirement. However, they are relatively expensive to fabricate. Examples include Sulzer Wire Gauze BX and CY, Koch Glitsch Goodloe, Montz A3-500, AMACS Woven Gauze. Figure 1 shows the photographs of some wire gauze packings.

2.2 Second Generation: Standard Structured Packings

Because of the high fabrication costs associated with the wire gauze packings, more cost-effective corrugated sheet metal structured packings were introduced in the late 1970's. These packings have since become the standard of the industry. Examples are shown in Figure 2.

Because these packings have a sharp change in angle between the layers, the vapor and the liquid have to change direction abruptly at the interface of the packing layers. Such a sudden change in the direction of the liquid and vapor flow can limit the performance of the packings.

2.3 Third Generation: High Capacity Structured Packings

The third generation, or high capacity structured packings, are very similar to the standard packings, except that they provide smooth transitions between layers for the liquid and vapor streams. Examples are shown in Figure 3.

3 Anatomy of a Corrugated Structured Packing

Structured packing comprises metal sheets that are textured, crimped, and often crimped – and assembled together as blocks.

3.1 Texture and Perforation

Texturing can be seen, for example, in the photograph of the Mellapak packing of Figure 2. Texturing promotes liquid film spreading while perforating minimizes vapor channeling and promotes vapor-liquid contacting. Texturing and perforating can increase the packing efficiency.

3.2 Blocks and Layers

Structured packings are usually built in layers of heights 200-300 mm, which are further segmented into multiple blocks. Figure 4 shows a full bottom layer of the packing with a few blocks of another layer placed on top of the bottom layer.

3.3 Crimps

The crimps are an important physical feature of a structured packing, and effectively determine the surface area and flow patterns. Referring again to Figure 2, the crimps of most of the structured packings are easily seen.

3.3.1 Crimp Angles

Based on the crimp angle from the vertical axis, structured packings are classified into X (30°), Y (45°), and Z (60°) packings. More-vertical crimps exhibit lesser hold-ups, lesser pressure drops, lower efficiencies, and higher capacities

3.3.2 Crimp Depth

Large crimp depths provide a low surface area, and conversely, small crimp depths provide a larger surface area. Obviously, to create more surface area, more metal is required - more metal means more cost, but also a higher efficiency. With more surface area, liquid hold-ups are higher, pressure drops are higher, and capacities are lower.

4 Installation of structured packings

First, the packing support grid is installed, and then, packing elements are lowered down by means of a belt or a rope to prevent damage to the packing. Metal guide plates or shoes are employed to facilitate installation of packing. The installer stands on a board to prevent damage to the packing sheets. As shown in Figure 4, each layer contains wall wiper bands that are bent out so they touch the column wall so that any liquid wall flow will be redirected back into the bulk of the packing. Each layer is usually rotated 90 degrees to minimize seam alignments of the packing blocks, and to prevent short-circuiting of vapor liquid traffic. Careful installation of the structured packing is vital for the column to work.

5 Performance Characteristics of Structured Packings

The key performance characteristics are HETP, Capacity or often, the Maximum Useful Capacity (MUC), and pressure drop.

5.1 Pressure Drop

Pressure drops of structured packings are lower than those of random packings, and much lower than those of trays. The pressure drop across a structured packing increases gradually with increasing liquid and vapor rates; the pressure drop increases sharply close to the flood point, as liquid begins to stack up in the bed.

5.2 HETP (Height Equivalent to a Theoretical Plate)

Structured packings use the concept of a "Height Equivalent to a Theoretical Plate", or HETP, to describe the performance of the packings. The HETP is defined as

$$\text{HETP}=\frac{\text{Packing Bed Depth}}{\text{Number of theoretical plates}} \tag{1}$$

The number of theoretical plates[1] required for a given separation is usually determined by a process simulation. An easy separation requires few theoretical plates; a difficult separation requires many theoretical plates.

For the same number of theoretical plates, a lower HETP results in a lesser packed bed depth, and consequently, a shorter column. Therefore, the lower the HETP, the more efficient the column is.

---

[1] Number of theoretical plates is also sometimes referred to as the number of theoretical stages, or the number of equilibrium stages.

For example, a highly efficient structured packing might achieve an HETP of about 300 mm (12 inches); a less efficient structured packing might achieve an HETP of 500 mm (20 inches).

5.3 Capacity or Maximum Useful Capacity (MUC)

The highest vapor rate at which the HETP of the column is good is called the MUC. A further increase in rate from the MUC will cause the HETP to rise rapidly, as the column progresses into a flooded condition.

5.4 Effect of packing specific surface area

Specific surface area, determined by the crimp characteristics, is an important factor affecting performance characteristics of structured packings.

High surface area structured packings exhibit high pressure drops and vice-versa. Typical pressure drop curves are shown in Figure 5.

High surface area structured packings exhibit low HETP's and vice-versa. Typical HETP curves are shown in Figure 6.

High surface area structured packings exhibit low MUC's and vice-versa. In Figure 6, the MUC point is indicated for both high surface area and low surface area packings. .

6 REFERENCES

Some of the best sources of information regarding structured packing are listed below:

Kister, H.Z., **Distillation Design**, McGraw-Hill, 1992

Kister, H.Z., **Distillation Operation**, McGraw-Hill, 1990

Perry's Chemical Engineer's Handbook, 8th edition, "Chapter 14 – Equipment for Distillation and Gas Absorption"

Strigle, R.F., **Packed Tower Design and Applications**, 2nd edition, Gulf Publishing Co., Houston, TX, 1987.

Vendor Resources:[2]

- http://www.sulzer.com/en/Products-and-Services/Separation-Technology/Structured-Packings
- http://www.koch-glitsch.com/masstransfer/pages/structured_packing.aspx
- http://www.amacs.com/packing/structured/
- http://amacs.com/wp-content/uploads/2012/09/AMACS-Montz-Brochure.pdf
- http://raschig.com/-Structured-Packings

[2] The listing here is not comprehensive. The list of references is provided for the benefit of the reader. FRI does not endorse or recommend any particular vendor product over another, including those listed in this document.

7 FIGURES

Figure 1: Wire Gauze Structured Packings

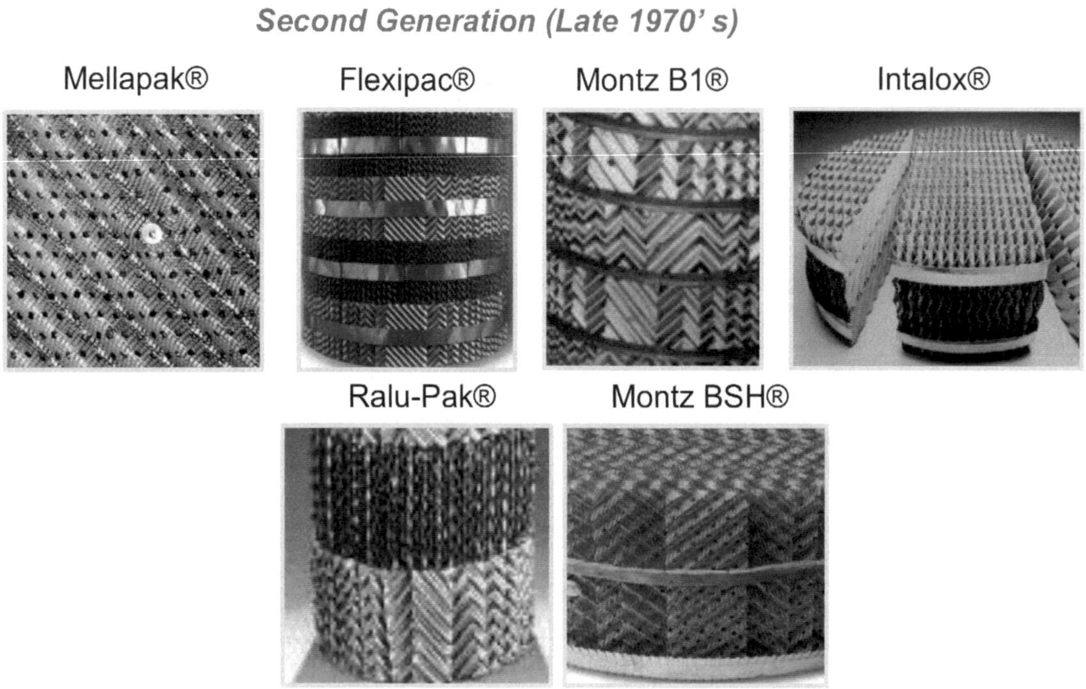

Figure 2: Standard Sheet Metal Structured Packings

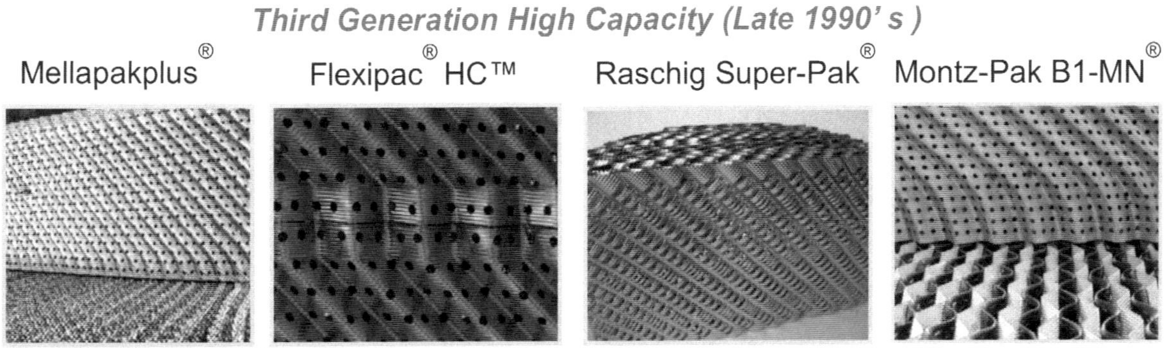

Figure 3: High Capacity Sheet Metal Structured Packings

Figure 4: Sulzer Mellapak 250.Y Packing - Packing Layer and Blocks

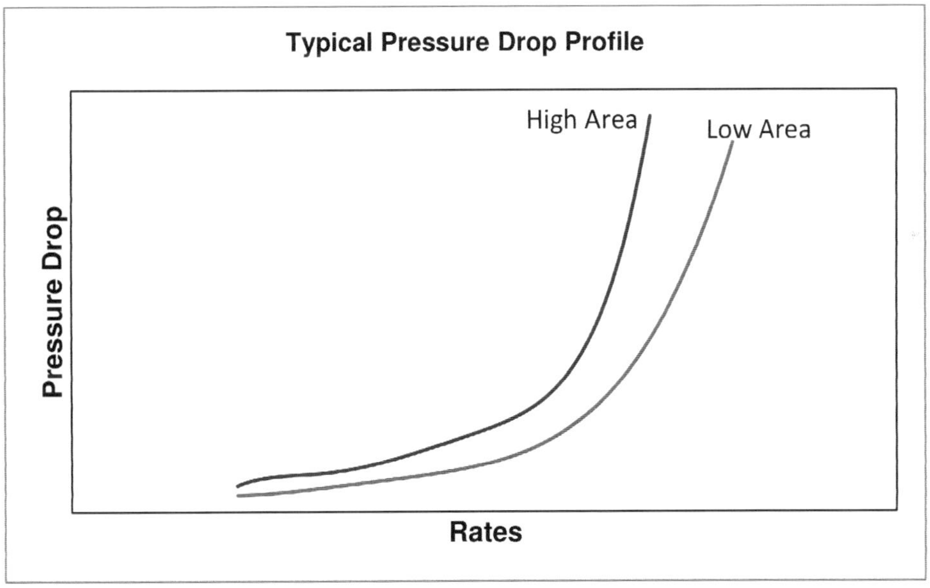

Figure 5: Typical Pressure Drop Curves

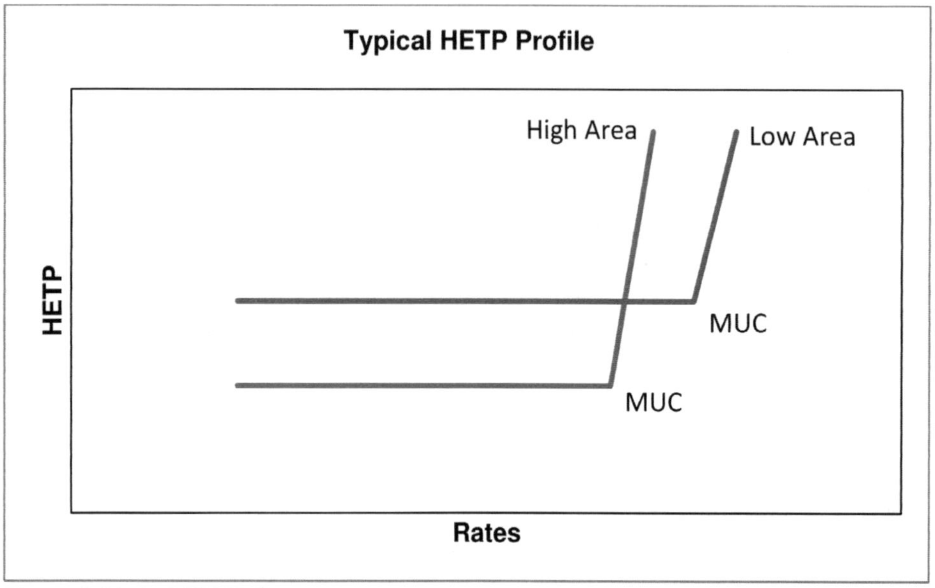

Figure 6: Typical HETP Curves

Liquid Extraction
Processes and Equipment

By

Mike Resetarits
AY Ogundeji
Tony Cai

Fractionation Research Inc.
424 S. Squires Street, Suite 200
Stillwater, OK 74074

Presented at YP Tutorials: Distillation and Absorption Equipment,
AIChE Spring Meeting, New Orleans, Louisiana, 2 April, 2014

Abstract

What is the best way to remove C_6, C_7 and C_8 compounds from a feed stream containing C_6, C_7 and C_8 compounds? How do you take a lubricating oil and remove certain compounds and improve its viscosity index (so that its viscosity does not change excessively with the oil's temperature)? What is propane deasphalting? The answers to all of these questions is "liquid extraction processes."

What technology was supported and funded by the actor Kevin Costner to help clean up the large 2010 Gulf of Mexico oil spill? What are Karr reciprocating plate columns, Oldshue-Rushton contactors and Podbielniak centrifuges? What the heck is a Graesser Raining Bucket? The answer to all of these questions is "liquid extraction equipment."

Imagine heavy and light fluids flowing vertically past each other with only a very small density difference. Imagine rotating equipment that works better without rotating. Liquid extraction is different – very different. There are very few experts.

Keywords: Extraction Aromatics Ternary Equilibria

Acknowledgment

The authors acknowledge the many contributions of Dr. Frank Seibert and Dr. Bruce Eldridge (both of the University of Texas at Austin) to the field of liquid-liquid extraction.

OVERVIEW

Some people say that chemical engineering is 50% Reactions and 50% Separations. In the field of Separations, distillation is the technique that is employed the most often, but liquid-liquid extraction, absorption, adsorption and membranes also play important roles.

Liquid-liquid extraction has many, and diverse, applications. Extraction devices can employ agitation to hasten mass transfer; there are many types of agitated devices. There are also many types of devices that do not employ agitation. Mass transfer calculations usually begin with liquid-liquid equilibrium (thermodynamic) data. Hydraulic calculations usually begin with force balances around droplets, and interfacial tension measurements.

GENERAL

As shown in **Figure 1**, when distillation is employed, a liquid phase and a vapor phase flow countercurrently. With extraction, a heavy liquid phase and a light liquid phase move countercurrently. With extraction, those two phases must be generally immiscible in each other. Chemical compounds in one phase, however, diffuse into the other phase. Very often, one liquid phase is aqueous; the other liquid phase is hydrocarbon. (Many people know that water and oil do not mix.) **Figure 2** shows a swarm of light liquid droplets (maybe "oil") rising through a heavier liquid (maybe "water"). At the top of the photograph, the light liquid droplets are seen to be coalescing.

Extraction columns usually have two feed streams. One is called the "Feed" and the other is called the "Solvent." The Feed stream of **Figure 3** is comprised of just two chemicals, A and B. The solvent stream is designated by "S". "A" is the "Solute" which ultimately transfers from the Feed stream to the solvent stream. After the Feed stream loses component A, the residual liquid, which is mostly component B, is often called the "Raffinate." (Generally, in the extraction world, the word "raffinate" refers to a low-value product stream.) In Figure 3, the product stream that is designated by "S + A" usually goes to a solvent recovery column that separates S from A. The resultant solvent stream is returned to the extractor. "A" is usually the final target product. The solvent recovery column is usually some form of distillation column, e.g., an extractive distillation column.

In trayed distillation and absorption columns, the vapor phase is dispersed as droplets in the downward-flowing liquid stream. As shown in Figure 3, extraction processes offer four different possibilities as follows:

1. Solvent = heavy phase; feed = light phase; heavy phase is dispersed
2. Solvent = heavy phase; feed = light phase; light phase is dispersed
3. Solvent = light phase; feed = heavy phase; heavy phase is dispersed
4. Solvent = light phase; feed = heavy phase; light phase is dispersed

Generally, the heavy phase is an aqueous phase and the light phase is a hydrocarbon phase. Most extraction experts believe that the more voluminous phase should be dispersed.

APPLICATIONS

A partial list of extraction applications is provided below:

- Pharmaceutical – production of penicillin
- Food – decaffeination of coffee
- Metals extraction
- Chemical and petrochemical
  - Aromatics recovery
  - Lube oil extraction
  - Acetic acid extraction
  - Desulfurization
- Liquid sweetening

BTX aromatics extraction is a particularly popular process, especially in refineries. A product stream from a catalytic reformer contains hexane, benzene, heptane, toluene, octane, xylenes and heavier aromatic and non-aromatic compounds. The aromatic (BTX) compounds are far more valuable than the non-aromatic compounds. The former cannot be separated easily from the latter via distillation columns.

Figure 4 shows a light liquid feed stream (C_6-C_8) entering the side of an extractor. A heavy liquid solvent stream enters the top of the extractor. There are at least two popular solvents – sulfolane and tetraethylene glycol. Both of these solvents have a strong tendency to absorb (or "extract") the benzene, toluene and xylenes, leaving behind the hexane, heptane and octane, i.e., the non-aromatic raffinate stream. Rich Solvent leaving the bottom of extractor goes to a (vapor-liquid) Solvent Recovery Column which separates the solvent from the BTX – the final product. That Recovery Column also generates a light reflux stream that is returned to the extractor. The BTX product stream goes to a series of distillation columns that separate and purify the benzene, toluene and xylenes.

Extraction can also be used to remove heavy aromatics compounds from a lubricating oil feed stream. The most popular solvents are furfural, phenol and NMP. The final product from the extractor is a lubricating oil that has an improved Viscosity Index, i.e., its viscosity is less dependent on temperature.

CONTACTORS

There are many types of liquid-liquid contacting devices. Some employ agitation, e.g., mixers, to hasten mass transfer. Others do not require or employ agitation. For those, in fact, agitation can be harmful, creating a totally-emulsified contactor wherein the light and heavy phases never separate out. The interface between any two liquids has an intermolecular "skin." The strength of that skin is called the "interfacial tension." Many extraction experts believe that liquid-liquid systems and interfacial tensions higher than 15 dynes/cm have a tendency towards large droplets and low interfacial areas for phase-to-phase mass transfer. Such systems require agitation to create more, smaller droplets. Liquid-liquid systems with interfacial tensions less than 15 dynes/cm have a tendency to easily produce many small droplets and a large interfacial mass transfer area. Agitation is not required with such systems; efficiencies are acceptable without it.

Generally, liquid phase diffusivities (diffusion coefficients) are much lower than vapor phase diffusivities. With a liquid-liquid extractor, diffusing solute compounds move slowly through one liquid phase, pass through the interface, and then move slowly through the second liquid phase. As a result of these slow mass transfers, extractor efficiencies are generally low. For example, a typical trayed extractor might have a tray efficiency of only 20%.

A partial list of agitated and non-agitated extraction devices is given below:

Agitated
- Mixer-settlers
- Podbielniak centrifuges
- Costner Centrifugal
- Rotating Disc Contactors
- Scheibel
- Karr reciprocating plate
- Kuhni

Non-agitated
- Packing
- Conventional trays
- Seal Deck and Rain Deck (UOP) trays
- Multiple Upcomer (UOP) trays

Figure 5 shows the UOP Multiple Upcomer (MU) tray. With this tray, the heavy phase is the dispersed phase. Each tray deck has many circular perforations which create a swarm of heavy phase droplets. As the droplets fall through the continuous light phase, mass is transferred. Those droplets coalesce on the tray below. The cycle is repeated again and again. As shown in Figure 5, the light liquid phase snakes its way up the column as the continuous phase.

There are many proprietary agitated extraction devices. Rotating disc contactors, such as the one shown in **Figure 6**, have often been used in Sulfolane aromatics extraction units. The Costner Centrifugal Extractor has been considered for off-shore oil-water separations. Podbielniak extractors are popular in pharmaceutical separations.

HYDRAULICS

For some liquid-liquid systems, at some temperatures, interfacial tensions can be measured easily with a ring tensiometer such as the one shown in **Figure 7**. Initially, the ring is lowered to the bottom of the containment cylinder. Then, the ring is pulled upwards, by a tin wire, until the interface between the two liquid phases is impacted. At that point, the interfacial "skin" resists any other upward ring movement. Eventually, the ring breaks through the skin with a force that is measured. That force is easily converted to an interfacial tension in dynes/cm. Thereafter, if the ring is pulled upwards through the light liquid phase, the liquid-vapor interface will soon be impacted. The force that will be required to pull the ring through that interface is a direct measure of the light liquid surface tension. For liquid-liquid systems at higher temperatures and pressures, interfacial tensions can be determined indirectly from the sizes of droplets that are created by sieve plates.

THERMODYNAMICS

Distillation calculations employ the concept of "equilibrium stages" to determine column heights. Vapor-liquid equilibrium laboratory data are employed to determine the number of equilibrium stages that are required to effect specified distillation separations.

Similarly, the extraction technology employs the concept of "equilibrium stages," and liquid-liquid equilibrium data, to determine staging requirements and contactor heights (sizes). Unlike distillation, extraction often employs ternary (triangular) diagrams to represent liquid-liquid equilibria. **Figure 8** provides a ternary diagram for the BTX example of Figure 4. The three apexes of the diagram represent pure components. Every position within the triangle represents a ternary composition. Every position under the curved umbrella represents a non-equilibrium mixture that sooner or later becomes two immiscible liquid phases that are in equilibrium with each other. The exact middle of the triangle represents a non-equilibrium mixture of 33% toluene and 33% heptane and 33% sulfolane. In a BTX extractor, Sulfolane is the solvent, toluene is the solute and heptane is the raffinate. Regarding this diagram, the feed stream is a toluene/heptane mixture.

In Figure 8, Points a, c and e represent light liquids that are high in heptane concentrations, low in toluene concentrations and almost absent of sulfolane. Points b, d and f represent heavy liquids that are high in sulfolane concentrations, high in toluene concentrations and almost absent of heptane. The almost-horizontal lines are called "tie lines." The points at the ends of each tie line represent light and heavy phases that are in equilibrium with each other. For example, Points a and b are in equilibrium with each other. The "Plait Point" represents the unique situation where the light and heavy liquid phases have the exact same compositions. Ternary diagram equilibrium data can be employed to determine the number of equilibrium phases that are required to effect target separations. Mass transfer models and computer programs exist for determining tray and packing efficiencies. More often than not, new extractors are designed using the efficiencies that were observed with similar extractors that are already operating.

REFERENCES

Some of the best sources of information regarding the science of liquid-liquid extraction are listed below:

- Treybal, R.E., **Liquid Extraction**, McGraw-Hill 1963.
- Lo, T.C., Baird, M.H.I., and C. Handson, **Handbook of Solvent Extraction**, John Wiley and Sons 1983.
- Laddha, G.S. and T.E. Degaleesan, **Transport Phenomena in Liquid Extraction**, McGraw-Hill 1978.
- Arlt, W., Macedo, M. Rasmussen, P., and J.M. Sorensen, **Liquid-Liquid Equilibrium Data Collection, vol. V, Parts 1-4,** DECHEMA Chemistry Data Series, Frankfurt 1997.
- **Perry's Chemical Engineer's Handbook**, 7th edition, "Liquid-Liquid Extraction Operations and Equipment", 1997. (revision in 2006)
- Walas, S.M., **Phase Equilibria in Chemical Engineering**, Butterworth Publishers 1985.
- Maddox, R.N., **Gas and Liquid Sweetening**, Campbell Petroleum Series, 1977.

FIGURES

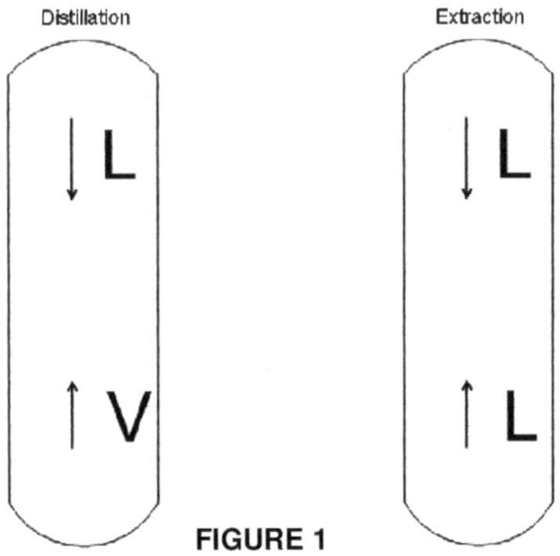

FIGURE 1

FIGURE 2

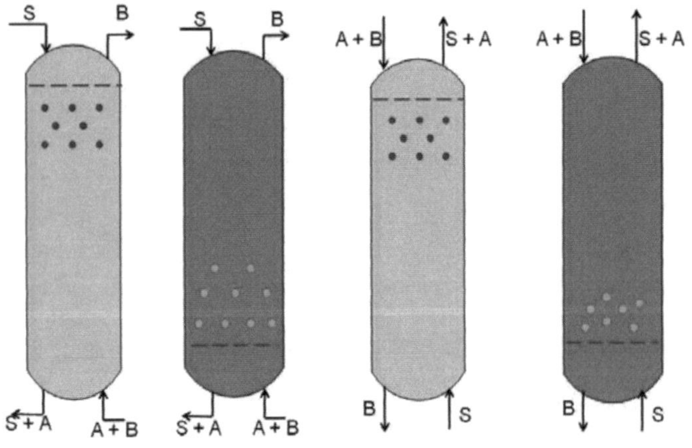

FIGURE 3. Extraction Options

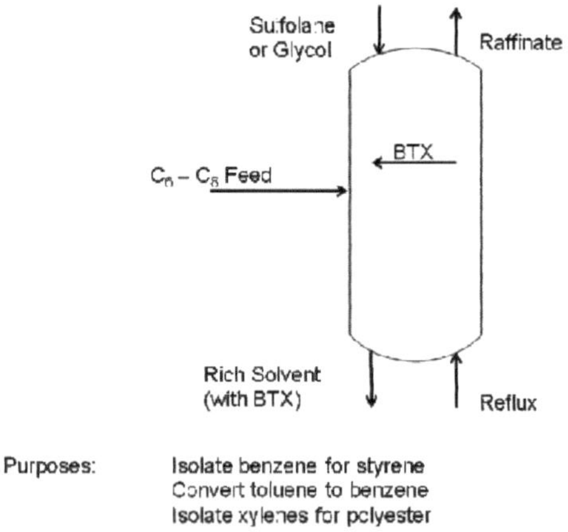

FIGURE 4. Aromatics Extraction

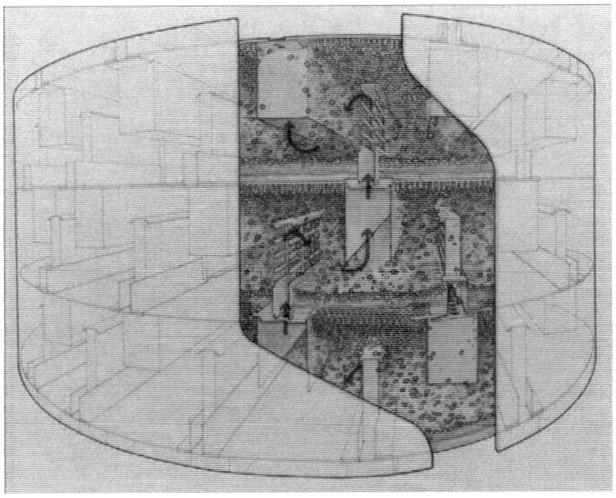

FIGURE 5. MU Tray Hydraulics

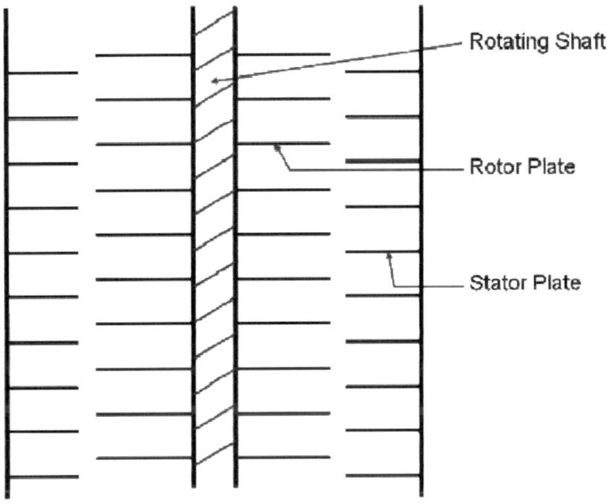

FIGURE 6. Rotating Disc Contactor

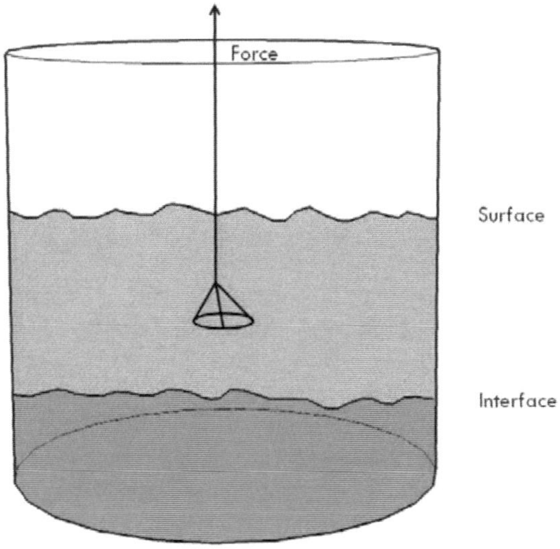

FIGURE 7. Ring Tensiometer

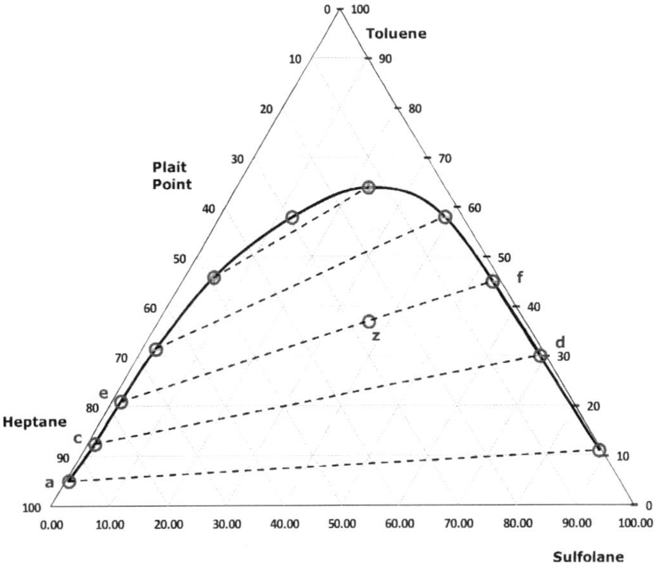

FIGURE 8. Heptane + toluene + sulfolane at 25°C, a type I system. [*Data taken from De Fre and Verhoeye,* J. Appl. Chem. Biotechnol., ***26,*** *pp. 1-19 (1976).*]

Acid Gas Absorption Fundamentals

By

Clint P. Aichele
James. R. Whiteley
Yash Tamhankar

School of Chemical Engineering
Oklahoma State University

Mike Resetarits

Fractionation Research Inc.
424 S. Squires Street, Suite 200
Stillwater, OK 74074

2 April 2014

Presented at YP Tutorials: Distillation and Absorption Equipment,
AIChE Spring Meeting, New Orleans, Louisiana, April 2014

ABSTRACT

Absorption is a standard unit operation that has been studied by chemical engineering students for decades. One of its primary uses is for the removal of acid gases, like CO_2 and H_2S, from natural gas streams. Historically, aqueous amines and hot carbonate solutions have been used to perform these absorptions. Many different amines, solution strengths, activators and passivators have been tested and employed industrially. Approximately half of the industrial towers are trayed; half, packed. Foaming adversely affects the hydraulic performance in these systems.

Acid gas absorption is receiving appreciable new attention as concerns rise regarding CO_2 in the atmosphere. In addition, the United States has recently experienced significant growth opportunities in shale gas production which require improvements in gas treating technologies. This paper reviews acid gas absorption and regeneration fundamentals. A PFD will be discussed at length, to illustrate important concepts. New absorbent solutions will be discussed, including ionic liquids. Flue gas applications will be discussed, including the difficulties associated with column pressure drops and very large column diameters. Rate-based modeling approaches will also be reviewed.

INTRODUCTION

Subsequent to the Industrial Revolution, global carbon dioxide emissions have increased from virtually zero to 9 billion total metric tons. These emissions originated from the natural gas, petroleum, coal and cement industries. Carbon dioxide, CO_2, emissions increase annually, as an increasing percentage of the global population consumes gas/oil/coal energy and drives cars. Carbon dioxide is a "greenhouse gas" and many people contend that such gases are contributing to the warming of the planet. As CO_2 emissions continue to increase, the technologies that reduce those emissions receive increasing attention from academia, industry and global governments.

For many years, and now, acid gas absorption units have been the primary means whereby CO_2 emissions have been reduced. Such units are – everywhere – and are relatively easy to understand. Several excellent books describe such units. [1,2,3]

DISCUSSION

Natural Gas Processing

When it first comes from the ground, natural gas contains many compounds, as shown in **Table 1**. Methane burns nicely in a furnace, but CO_2 does not burn at all. Therefore, the consumers who purchase natural gas do not want to receive, and pay for, CO_2. Typically, natural gas production companies must reduce the CO_2 content of natural gas down to 2% or less.

Raw natural gas sometimes contains hydrogen sulfide, H_2S, which is particularly-dangerous poison. Natural gas production companies must reduce the H_2S content of natural gas down to 4 ppmv or less.

Very recently hydraulic fracturing or "fracking" is yielding gases that are very high in ethane, propane and butane. These "natural gas liquids" or "NGL's" are very valuable in that they can easily be converted to ethylene and propylene. As fracking increases in the U.S., more raw natural gas streams are being created and all of those streams need the acid gases (CO_2 and H_2S) to be removed.

Flue Gases

"Flue gas" often refers to a combustion exhaust gas from the chimney of a power plant. Its composition depends on what is being burned, e.g., natural gas or coal. **Table 2** provides typical flue gas compositions. Globally, the CO_2 content of flue gases is receiving increasing attention, but the removal of that CO_2 is particularly challenging. Flue gases are at very low (near atmosphere) pressure and need to be compressed to flow (through absorbers). Many flues create huge volumes of such gases and those gases contain particulates and oxygen. Perhaps the largest challenge of all: What can be done with the CO_2 once we capture it? "Sequestration" is an umbrella word that includes all of those natural and unnatural places where CO_2 can be stored or utilized. For example, compressed CO_2 can be used for Enhanced Oil Recovery (EOR).

Solvents

Water can be used as a solvent for the acid gases in a natural gas stream, but water is very ineffective. The Benfield hot carbonate process employs aqueous potassium carbonate as the solvent. CO_2 and H_2S are removed from natural gas streams via the following overall chemical reactions:

$$K_2CO_3 + CO_2 + H_2O = 2KHCO_3$$

$$K_2CO_3 + H_2S = KHS + KHCO_3$$

Globally, for acid and gas removals, aqueous amine solutions are probably the most popular. Forty years ago, MEA and DEA were the workhorses. Then, MDEA became popular especially because of its ability to remove H_2S quickly and totally, without over-absorbing CO_2. Most recently, piperazine is being employed regularly. **Figure 1** is a partial list and classification of amines. When CO_2 is absorbed into an aqueous amine solution, many chemical reactions are involved. These chemical reactions are very exothermic. Amine solutions are very corrosive. Passivators are sometimes employed to reduce attacks on the metals of acid gas processing units.

Typical Unit

Figure 2 shows a typical amine absorption unit. Two processing columns are employed – an absorber and a solvent regenerator.

Referring to **Figure 2**, the Feed Gas enters the bottom of the absorber. Typical pressures of that gas range from 100 to 1200 psia. Absorber shells can be as thick as 3 inches. As the gas proceeds upwards through the column, it encounters a down-flowing liquid solvent stream. Lean liquid solvent enters the top of the absorber. This lean solvent stream must be low in acid gases (CO_2 and H_2S) or gases issuing from the lean solvent stream will contaminate the gas product stream. It is also important that the lean solvent stream be relatively cool to remove CO_2 to the lowest possible levels. Additionally, a too-warm lean solvent will flash off its acid gases too easily. The treated gas product stream that leaves the top of the absorber often goes to a train of distillation columns that separates and purifies the methane, ethane, propane, isobutane and n-butane.

Still referring to Figure 2, as the liquid solvent stream flows down through the absorber, the exothermic reaction associated with the chemical absorption causes the solvent to increase appreciably in temperature. Most of the temperature increase occurs near the bottom of the absorber. After the rich solvent leaves the bottom of the absorber, it often goes to a heat exchanger that increases the rich solvent temperature further. The hot rich solvent goes to the regenerator, where a massive flash occurs especially because the regenerator operates at near-atmospheric pressure. Some regenerators have two in-series flash chambers in their tops to allow additional flashing off of the acid gases. All of the vapors leaving the top of the regenerator (steam, CO_2 and H_2S) go to a condenser that liquefies the steam and returns the water to the top of the regenerator.

After flashing, the rich solvent flows all the way down the regenerator until it reaches a reboiler. That heat exchanger boils off a large portion of the water that is in the solvent. Very little amine is boiled off. The resultant steam flows upwards through the regenerator, and strips the acid gases out of the down-flowing solvent.

The lean solvent that leaves the bottom of the regenerator is very hot (depending on the reboiler temperature) and is low in pressure. A pump is required to increase the pressure of the lean solvent to a pressure somewhat higher than the absorber pressure.

A high-temperature lean solvent would be very ineffective at acid gas removal, especially due to vapor-liquid equilibrium (thermodynamic) considerations. The rich solvent can be used to decrease the temperature the lean solvent. The lean solvent temperature can also be reduced using a cooling water exchanger, as shown in Figure 2.

There are many variations of absorption unit processing schemes. Examples follow:

1. A semi-lean solution can be pulled from the side of the regenerator and fed to the side of the absorber
2. The bulk of the lean solution can be fed to the side of the absorber, with only a slip stream of the lean solution going to the cooling water exchanger, and the top of the absorber

Regarding the acid gases that leave the top of the regenerator, the H_2S can be converted to elemental sulfur in a Claus Unit. Historically, the CO_2 has been released into the atmosphere. It can, however, be compressed and utilized for Enhanced Oil Recovery (EOR).

Other Solvents

Aqueous amines and potassium carbonate solutions have historically been the most commonly employed absorption solvents. Some amine solutions contain a mixture of primary, secondary and/or tertiary amines. Some potassium carbonate solutions employ amines as "activators." Generally speaking, H_2S is absorbed quickly into amine solutions; CO_2 slowly. When an excess of CO_2 is absorbed by the lean solution, an excess of regeneration (and energy) is required. A perfect amine solution formulation would simultaneously exactly achieve the targeted CO_2 and H_2S product gas compositions.

Solvents other than aqueous amines and hot carbonate solutions have been employed and are being investigated

Physical solvents, ionic liquids, and enzymes have been explored as possible CO_2 capture solvents to replace aqueous amine absorption for flue gas applications(4). These approaches have been investigated primarily for their ability to reduce the size and overall cost of acid gas absorption applications. Physical solvents require the physical solubility of the acid gas into the solvent; therefore, physical solvents are commonly used for high pressure applications such as synthesis gas scrubbing or natural gas absorption. Because physical solvent processes don't rely on chemical reactions to absorb the acid gases, the solvent regeneration energy is typically less than that associated with conventional aqueous amine absorption. Ionic liquids have received considerable attention for their potential ability to reduce the regeneration energy of the CO_2 removal process(5). However, their large-scale use for CO_2 removal applications has not been achieved. Enzymes show promise as a way to enhance the reaction rate of CO_2 into solution, thereby reducing the size of the absorber. However, large scale demonstration of this technology at competitive costs remains to be seen. Despite these advances in solvent development, aqueous amine absorption remains the most competitive technology for CO_2 removal from flue gas, and it continues to be a technology of choice for natural gas treating and refinery scrubbing applications.

Column Internals

Historically, most hot carbonate columns have employed packings, especially random packings. For many years, most amine units have employed 20 trays in the absorbers and 20 trays in the regenerators. In recent years, however, packings have sometimes been employed in amine columns, especially when the lower pressure drops of packings have yielded processing benefits.

Foaming

Acid gas absorbers are known "foamers," i.e., the 2-phase mixtures on the trays or in the packed beds bubble excessively. These foams cause column capacities to be reduced. FRI has demonstrated that packings do not break foams, but the impact of the foaming is less with packings than it is with trays. On trays, foaming can be reduced, somewhat, via short froths, large deck holes and designing nearer the spray regime (as compared to the froth regime)(6) .

With time, aqueous amine solutions degrade and begin to carry particulates. It is generally held that these particulates cause foaming to increase. With operating untis, old amine solutions need to be filtered – and augmented regularly with fresh amine solution.

Equilibrium Stages and Process Simulations

With amine units and hot carbonate units, the absorber trays/packings and the regenerator trays/packings do not need to yield very many equilibrium stages (or "theoretical trays") to achieve the targeted absorption and regeneration separations.

McCabe-Thiele graphical constructions show huge areas between the operating and equilibrium curves. Step-offs between those curves show that a typical absorber only requires 1-3 equilibrium stages; likewise for a typical regenerator. At the very same time, in these mass transfer services, tray and packing efficiencies are low and difficult to calculate because of the chemical reactions and equilibria that are involved. Typical tray efficiencies are around 20%; some are as low as 5%. Because of these unusual circumstances and difficulties, rate-based theoretical calculations are increasingly preferred for unit simulations. For example, the ProTreat computer program of Optimized Gas Treating[7] is very versatile and can be used to calculate/predict column performances without using the concepts of equilibrium stages or tray/packing efficiencies.

Safety Considerations

Amine solutions are very corrosive. They can dissolve metal and burn skin. Acid gas absorbers are operated at very-high pressures, as high as 1200 psi. Pressure relief systems are, of course, required on such units.

In 1984, welding work was performed inside an amine absorber in Illinois. Following that work, Post-Weld Heat Treatment (PWHT) work was not performed. A weld failed due to stress corrosion cracking. The gas that escaped from the operating column ignited. The column ended up landing 1 kilometer away. Seventeen workers were killed. There was approximately $100 million in property damage. All pressure vessels must be properly designed and maintained.

Current Research

Acid gas absorption/regeneration is a very popular research area in 2014. At the November 2013 AIChE meeting in San Francisco, there were 22 sessions with "CO_2" or "carbon dioxide in their titles. Those sessions included 130 presentations! The presenters represented governments, academia and industry.

Many aspects of acid gas treating are now being studied globally. Two examples follow:

1. Dr. David Van Wagener (Ph.D. from the University of Texas with Dr. Gary Rochelle) focused on 2-stage flashing to reduce regeneration energy consumption[8].
2. Oklahoma State University is studying spray columns for the reduction of the capital cost associated with flue gas treating (see **Figure 3**)[9].

There are many other examples of on-going research work.

CLOSING

Acid gas removal from gas streams is not new. For many years, global gas plants have been removing CO_2 and H_2S from natural gas streams, especially using aqueous amines and hot carbonate solutions.

During recent years, acid gas removal has received appreciable fresh attention. Planet Earth is coming out of a relatively recent ice age, and so, it is generally held that Earth is warming. Many people believe that the warming is being exacerbated by human actions, especially the combustions associated with automobiles and power plants. Like the metabolism reaction, the combustion reaction yields CO_2 as a product. CO_2 is regarded as a "medium" strength" greenhouse gas. The reduction of global CO_2 emissions is correctly on many people's minds, these days, and in many scientists' laboratories.

REFERENCES

1. Maddox, R.N., Gas and Liquid Sweetening, 2nd printing, John M. Campbell, Norman, OK, 1977.
2. Humphrey, J.L. and Keller II, G.E., Separation Process Technology, McGraw-Hill, New York, NY, 1997.
3. Seader, J.D., Henly, E.J. and Roper, D.K., Separation Process Principles: Chemical and Biochemical Operations, 3rd Ed., John Wiley & Sons, Inc., Hoboken, NJ, 2011.
4. Figueroa, J.D., Fout, T., Plasynski, S., McIlvried, H., Srivastava, R.D., "Advances in CO_2 Capture Technology – The U.S. Department of Energy's Carbon Sequestration Program," International Journal of Greenhouse Gas Control, 2, 9-20, 2008.
5. Brennecke, J.D. and Gurkan, B.E., "Ionic Liquids for CO_2 Capture and Emission Reduction," The Journal of Physical Chemistry Letters, 1, 3459-3464, 2010.
6. Resetarits, M.R., Navarre, J.L., Monkelbaan, D.R., Hangx, G.W.A., and van den Akker, R.M.A., "Trays Inhibit Foaming", Hydrocarbon Processing, March 1992, p. 61-64.
7. Weiland, R.H. and Hatcher, N.A., Advanced Gas Treating: The Engineering Science, Optimized Gas Treating Inc., Houston, TX, 2011.
8. Van Wagener, D.H., "Stripper Modeling for CO_2 Removal by Monoethanolamine and Piperazine Solvents," PhD thesis, Univ. Texas Austin, 2011.
9. Tamhankar, Y., Whiteley, J.R., Resetarits, M., and Aichele, C.P., "Spray Droplet Characterization Inside a Glass Column Through Dense Wall Flow," Atomization and Sprays, 24, 115-128, 2014.

Table 1
Typical Natural Gas Concentrations

Gas	Feed	Product Specification
CH_4	70-95 %	-
C2-C4 (NGLs)	0-20 %	-
CO_2 (no heating value)	0-20 %	< 2 % for pipeline, < ~100 ppmv for cryogenic
H_2S (lethal)	0-5 %	< 4 ppmv
O_2	0-1 %	0.4 %
N_2	0-5 %	-
Ar, He, Ne, Xe	trace	-
H_2O	Equilibrium	4-7 lb/MMscf

Table 2
Typical Flue Gas Concentrations

Gas	Feed	Product Specification
CO_2	10-15 %	-
O_2	0-5 %	< 10 ppmv
N_2	70-75 %	-
H_2O	Saturated	< 1 %
SOx	~50 ppmv	-
NOx	~25 ppmv	-

Figure 1
Amine Chemistry

Monoethanolamine (MEA)

Primary amine
Fast reaction rates
Limited capacity

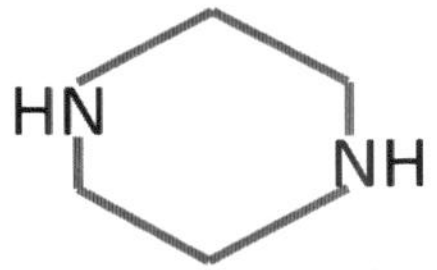

Piperazine (PZ)

Secondary amine
Faster reaction rates
Enhanced capacity

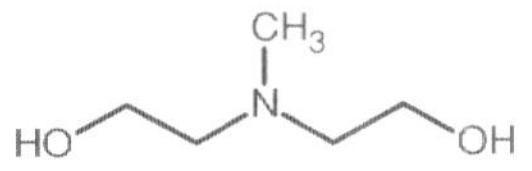

Methyldiethanolamine (MDEA)

Tertiary amine
Slow reaction rates
No carbamate formation
High capacity

Figure 2
Amine Absorption

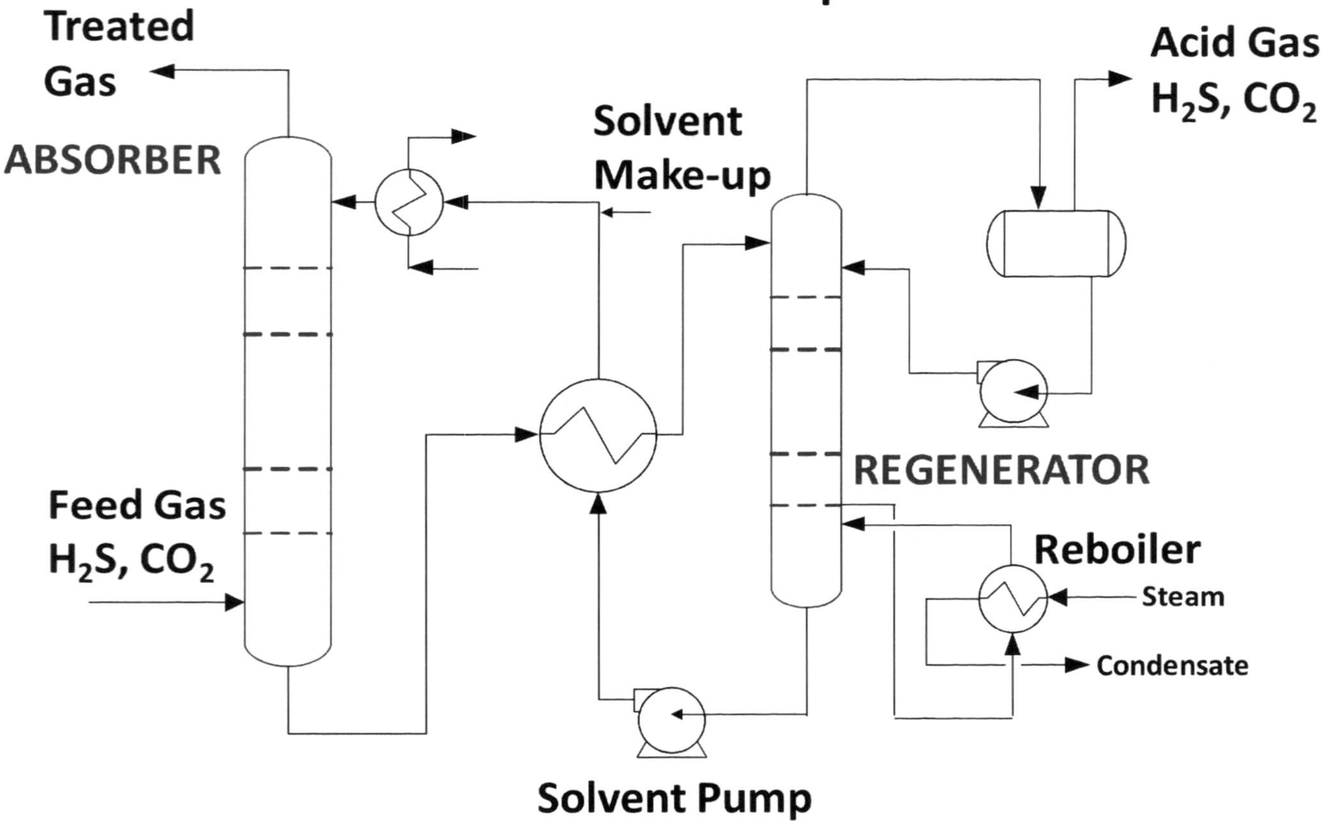

Figure 3
Oklahoma State University
Acid Gas Absorption Apparatus

Additional derivatives from gamma scanning of detecting flow distribution patterns will be demonstrated. The CAT-

350235 Radioisotope Applications for Obtaining Detailed Process

Abstract:

In all chemical plants and refineries there generally is not enough instrumentation in the right places in our processes. During the design phase, decisions were made between knowing more about the process, which means buying more instruments, and saving capital funds. When a problem appears with the operation of a vessel, there are always gaps in the knowledge of what is happening due to a lack of instruments in the places where knowledge is not routinely required. Process engineers and operations personnel need to know what is happening so they can take corrective action. When knowledge is missing, they often take actions that make matters worse.

For distillation and separation processes Chemical Engineers have historically relied upon plant process measurements such as flow rates, temperatures, and pressures and model or simulation results for data with which to solve troubleshooting or re-design problems. An array of on-line diagnostic services are available that can provide additional data offering real-time information on how pieces of process equipment are actually operating.

The most common application is Gamma Scanning. This test is primarily applied, but not necessarily limited to, distillation or separation columns. Gamma scans provide a density profile of the internal process of operating distillation columns and other process vessels. The density profile can be used to diagnose the hydraulic operating conditions of mass transfer devices such as damage to internals, flooding, degree of entrainment or weeping, liquid levels on trays and distributors, liquid distribution through packed beds, etc. The presenter will show case studies where scanning revealed vital process information that helped solve an operating problem, or helped make a revamp/re-design successful.

Scan technique is a specialized horizontal gamma scan used to generate a topographic profile of the internal cross-sectional density of process equipment. This profile is useful for the detailed study of liquid flow distribution through packed columns as well as having applications beyond distillation columns. A hybrid application involving tracers and scanning on fixed bed catalytic reactors will also be discussed.

Another technique involving radioactive materials is the Tracer Test, in which a radioactive tracer is introduced into the process for investigating the process flows. External radiation detectors are placed at strategically chosen locations outside vessels and piping to monitor the tracer flow through the process. Either liquid, gas or a solid tracer can be utilized depending on process conditions and the fluid to be studied. This tracer technique can identify liquid or gas maldistribution in fixed or fluidized bed reactors. The residence time through a tank or reactor can be measured and compared with the calculated residence time. Observed differences can relate directly to the degree of mixing or channeling inside the reactor. These tracer techniques have also been used in identifying leakage of heat exchangers and valves.

351759 Enhancing the Filtration Rate of Phosphogypsum Using Polymers

Abstract:

Phosphoric acid Phosphoric acid is an important intermediate product for production of fertilizers; it is produced either by wet process or by thermal process. The wet process using sulfuric acid is the most comment used process, in wish the phosphate rock ([fluor-apatite $Ca_{10}F_2\ (PO_4\ )_6$] is dried, crushed, and then continuously fed into the reactor and reacted (digested) with sulfuric acid (H2SO4). The reaction led to the production of phosphoric acid (H3PO4) and calcium sulfate dihydrate or phosphogypsum (CaSO4·2H2O).Phosphogypsum is separated from the reaction solution by filtration.

Filtration is a key operation in the manufacturing process of phosphoric acid, the objective of this operation is to separate the phosphogypsum and phosphoric acid. My presentation will be subdevised on two main parts; in the first part I will discuss the parameters that affect the filtration operation for the separation of phosphoric acid from phosphogypsum cake. In the second part, I will present a result of study for the filtration rate improvement and the reduction of P2O5 losses in the gypsum cake by using a suitable polymer. The results of this study are based on the online measurement of filtration rate and soluble content of P2O5 in gypsum cake, using bench scale of filtration unit. The study have been conducted in the case of two different process for phosphoric acid production Rhone Poulenc and Prayon.

352424 Simulation of Polymer Barriers with Various Inorganic Particulates to Achieve Ultra-Low Permeance of Small Molecules

Abstract:

The development of submicron membrane for filtration requires understanding of the movements of fluids both in the bulk regime and the molecular regime. Much of this same understanding is needed in creating barriers to small molecule diffusion. To this end, models to describe the effect of incorporating impermeable fillers on the transmission rate of small molecules through polymer barriers were developed. Two different types of particulates, spheres and platelets, of various sizes, arrangement and loading were investigated. Three dimensional simulation of diffusion through the structures were studied to provide deeper insights into the impact of these fillers. These simulation models looked at both the bulk diffusion and the Knudsen diffusion to capture the behavior for a large range of filler loads and for a large range of permeance for the base polymer.

Depending on the loadings of the filling and the concentrations of the diffusing molecule, the dominant mechanism for diffusion can vary between a bulk diffusion regime and a Knudsen diffusion regime. The various equations developed from the correlation in this work can be applied in both of these regimes. This allows for a better understanding of the impact of these fillers, especially, for the requirement of high barrier properties where both regimes could be important. The simulation approach used in this work can be employed to greatly reduce the amount of experimentation which is needed to develop novel barrier technologies and provide a methodology for future work in this area.

EXTRACTION OF BITUMEN FROM ATHABASCA OIL SAND BY COMBINING THE BENEFITS OF MECHANICAL SOLID LIQUID SEPARATION AND DRYING

Elisabeth Schmidt, Technische Universität Bergakademie Freiberg, Germany
Bernd Wacker, Siemens AG, Erlangen, Germany
Professor Dr.- Ing. Urs A. Peuker, Technische Universität Bergakademie Freiberg, Germany

Abstract

Canadian oil sand gained from open pit mining is processed with a hot water extraction to win the bitumen for later upgrading to synthetic crude oil. The problems arising with this process reach from high fresh water consumption over stable colloidal suspension, the so called Mature Fine Tailings, which form due to the presence of surfactants and are ecologically highly problematic. These facts formed motivations for an alternative process. Aims are the minimization of water usage, avoiding the formation of stable tailings by usage of organic solvents, maximizing the bitumen recovery and gaining a disposable solid, free from organic components. The alternative process unites a cake formation for filtration, a graded washing step and a steam pressure filtration application. The used organic solvents are toluene and n-heptane. With this process bitumen recovery rates up to 95 % can be reached even with a washing ratio of one. The residual humidity of water in the filter cake after steam impinging lies between 9 and 25 %. The ambition is to achieve a maximum of bitumen recovery and an organic and solvent free, disposable solid is reached.

Introduction

Canadian oil sand from open pit mining is processed by the *Clark Hot Water Extraction* (CHWE). The process has the following steps: mining, conditioning and transport, raw slurry production, bitumen extraction, tailings treatment, froth treatment, processing and upgrading of bitumen. The synthetic crude oil (SCO) production amounts to 0.4 million m^3/d [1], 60 % of which are open pit mined [2]. For this production 3 Mt of oil sand are mined daily. As to the CHWE the water usage amounts from 0.7 to 1.4 million m^3/d, meaning for every cubic meter of oil between three and six cubic meters water is needed. The fresh water is taken from the Athabasca River. Besides the high fresh water amount needed, the production of tailings and the high energy demand are further critical points. Main part of the energy is needed for heating up the process water. For the production of one cubic meter SCO 28 cubic meter natural gas are needed [3]. The average bitumen recovery is a total of 80 %.

Tailings of the CHWE can contain the so called Mature Fine Tailings (MFTs). This type of tailings forms due to the presence of hydrophobic clay particles in the oil sand and the usage of carbon hydroxide in water a stable colloidal suspension. The estimated time for sedimentation is 100 years. Different treatments are chosen before the tailings are induced into tailings ponds [4]. The area of the ponds is around 180 km^2 [5, 6]. 250 million l/d are induced into the tailings ponds. The critical facts of these ponds are the leakage into ground water, which is hard to evaluate (around 11 million l/d), the long consolidation time, the water containing toxic polycyclic hydrocarbons, naphthenic acids, heavy metals and bitumen particles floating on the surface. In summary the environmental and health risks are high.

The washing integrated steam pressure filtration

The essential aims for this alternative process arise from the deficits of the hot water extraction process:

- entire or maximum avoidance of water usage
- exclusion of the need of tailings ponds
- constant or increased bitumen recovery rates
- maximum solvent free, dry and disposable solids
- competitive energy demands

The process developed includes a filter cake formation, filter cake washing with changing washing fluid composition and a final steam application on the filter cake.

Materials and Methods

Oil sand is a mixture of sand, bitumen, water and clays to be found in different compositions all over the world. As the Canadian oil sand is the one with the highest mining rates worldwide and the development of the alternative processing is aimed on this mining area, the used oil sand originate from the Athabasca region. For the studies the oil sands were split in three quality classes. The bitumen content is given in percentage by weight. A poor quality oil sand contains less than 7 % of bitumen, a medium 12.8 %and a rich quality up to 15 % of bitumen. The poor and rich qualities are extremes in mining. The processed oil sand is blended in open pit mining, so the average bitumen content is 10 %. This guarantees stable conditions in processing.

Depending on quality the amount of fines in the oil sand varies. This material parameter is a critical one in the CHWE, as it affects the formation of stable colloidal suspensions. Bitumen droplets are entrained in the tailings, negatively affecting the bitumen recovery rates. The fines include particles smaller than 44 µm. In the used oil sand the following fines amounts are found: rich – 15 % by mass, medium – 26 %, poor – 40 %. The water content of the used oil sands is between 0.03 and 3 %.

Suspending and Filtration

The oil sand to be processed has to be suspended in a mixture of the organic solvents toluene and n-heptane [7-9]. By solving the carbon hydroxides from the solid phase, separation of the two phases is possible. The toluene is an aromatic solvent, solving the asphaltenes of the bitumen [10]. The asphaltenes are mainly condensed aromatic compounds linked by aliphatic chains. Due to their extremely high molecular mass the asphaltenes cause the high viscosity and density of the bitumen in the oil sand. The viscosity is linked to the bitumen recovery [11]. By solving the asphaltenes with the toluene the bitumen recovery can be maximized. If the asphaltenes remain unsolved, the bitumen viscosity is higher [12], as aggregation to high viscous particles can occur [13]. The second solvent is the aliphatic n-heptane. It cannot solve asphaltenes, but shows a smaller steam pressure than toluene, providing an advantage in the final steam application on the filter cake. Another advantage is the further reduction of the viscosity of the liquid phase.

The oil sand suspension is carried out in an optimized order, first toluene, secondly n-heptane. A simultaneous suspending leads to worse solving of the asphaltene flocculants. Suspending time is set at 1 minute, the shortest possible time for gaining a homogeneous suspension without clumps.

The filter cake formation is performed with a pressure nutsch filter at Δp=1 bar according to VDI guideline 2762. The filtration area is 19.6 cm². Filter media are adjusted to the experiments aims. Materials are E-CTFE or PP, pore widths vary between 6 an 27 µm. The solids volume concentration

c_V is varied between 0.21 and 0.33. By recording the filtrate mass during the filter cake formation a *t/V-V*-diagram can be gained and hence the filter cake resistance r_c and filter medium resistance R_m can be determined (see equation 1 and 2).

$$\frac{\Delta t}{V} = \frac{\eta_L \cdot \kappa \cdot r_c}{2 \cdot A^2 \cdot \Delta p} V + \frac{R_M \cdot \eta_L}{\Delta p \cdot A} \quad 1$$

$$\text{with } \kappa = \frac{c_V}{1 - c_V - \varepsilon} \text{ and } c_V = \frac{V_S}{V_S + V_L} \text{ with } V_L = V_{Sol} + V_{Bit} \quad 2$$

Based on a laminar flow the filter cake resistance equation accounts from the Kozeny-Carman equation [14]. The permeability coefficient k resulting from Darcy's law shows how the bulk structure influences the fluid flow through the pores (see equation 3).

$$k = \frac{\varepsilon^3}{5 \cdot S_V^2 \cdot (1-\varepsilon)^2} = \frac{\varepsilon^3 \cdot x^2}{180 \cdot (1-\varepsilon)^2} \quad 3$$

The pressure filtration is continued until the liquid surface reaches the filter cake's upper border. The saturation equals one at this point, meaning filter cake is thoroughly filled with pore liquid (see equation4). For the subsequent washing a complete wetting of the filter cake is given.

$$S = \frac{V_P}{\varepsilon \cdot V_{FC}} \quad 4$$

Filter Cake Washing

The washing step following the filter cake formation increases the bitumen recovery and creates the conditions for an optimal steam application. The washing fluids toluene and n-heptane are applied on the filter cake according to a fix washing ratio W, as a characteristic parameter for filter cake washing. The washing ratio specifies the volume of pore liquid V_P being replaced with unloaded washing liquid (see equation 5).

$$W = \frac{V_W}{V_P \cdot S} \quad 5$$

A shift in the density of the liquids during washing suggests the use of the mass relied washing ratio W_m (see equation 6). The mass of the washing liquid m_W is referred to the mass of the filter cake pore liquid m_P at saturation equal one.

$$W_m = \frac{m_W}{m_P} \quad 6$$

Analyzing a washing curve helps evaluating the needed amount of washing liquid for reaching a certain bitumen recovery. The bitumen concentration within the washing filtrate is the crucial parameter for gaining the dimensionless concentration ratio c/c_0, which is plotted over the washing ratio. The concentration ratio is gained by relating the current bitumen concentration in the washing filtrate to the bitumen concentration in the pore liquid at the start of the washing (see equation 7).

$$\frac{c}{c_0} = \frac{c_{Bit,W}}{c_{Bit,P,0}} \qquad 7$$

In ideal displacement washing the washing liquid will travel equally (piston-like) trough the cake [15-17]. Real effects lead to a quicker permeation of the washing liquid through larger pores and velocity gradients across the pore radius. Thus the so called fingering can occur [18]. If washing and pore liquid are miscible a mixing zone is formed, leading to a deviation from the ideal washing curve (see figure 1). The intermediate stage is reached, when c/c_0 drops below one. This stage mostly extends over a washing ratio of 0.5 to 1.5. In the last stage the solved bitumen diffuses from the interstitial particle liquid in the washing liquid flow. The concentration differences detected in the diffusion regime are small. The determination of the bitumen content within the washing filtrate is realized by mass balances. To reproduce the concentration curve during washing, the washing filtrate is divided in classes according to the washing ratio and analyzed separately. The bitumen concentration within the filter cake cannot be analyzed on-line; therefore the washing needs to be interrupted. A sample removal changes the cake structure, its volume and weight, preventing a comparable continuation of the washing. For plotting the concentration ratio of bitumen within the filter cake over washing ratio several experiments, stopped at different washing ratios, have to be compared.

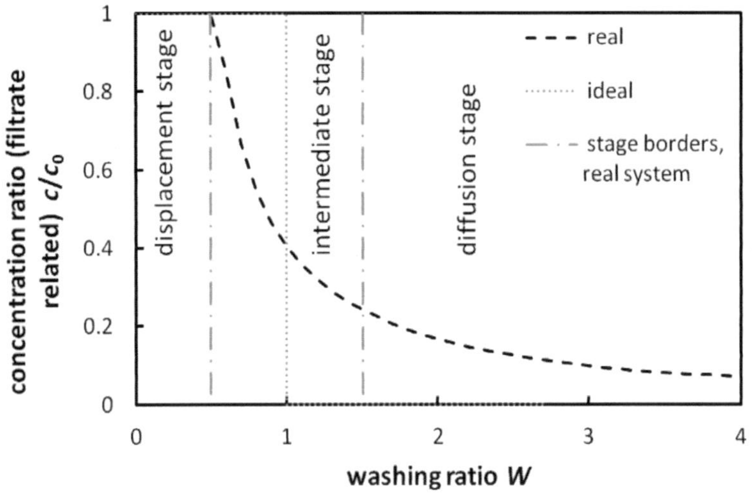

Figure 1. Real and ideal washing curves plotted as the dimensionless concentration of bitumen in the washing filtrate.

Washing is drawn out with gradually changing the composition of washing liquids – called graded washing. At small washing ratios (up to 0.33) toluene is present in the washing liquid to ensure an entire solving of the asphaltenes in the filter cake. With a progressing washing the n-heptane part in the washing liquid is increased to benefit from its lower boiling point in the subsequent steam application. The aim of the experimental is a washing ratio (set a 1) as small as possible with a maximum bitumen recovery. After the washing step the filter cake pores are filled with organic washing liquid with minimized bitumen content.

Steam application and drying

The processing of the oil sand is finished with mechanical-thermal dehumidification by application of steam and dry air. The cake formation unit used for the earlier steps is installed on a steam pressure nutsch. Thermo elements set on different levels of the cake formation unit allow the determination of the steam breakthrough. When saturated steam is applied on the filter cake a condensation layer forms on the cold cake surface (see figure 2). The condensed steam is used as a pressure medium for dehumidification with a differential pressure [19]. The condensation layer is macroscopic even and stable. The filter cake is slowly heated to the temperature of the saturated steam temperature by condensation heat. The sensible heat the filter cake gains can be calculated with the temperature difference, directly leading to specific condensate mass (see equation 8 and 9). The residual moisture consists of the condensed steam; organic content from washing liquid is negligible.

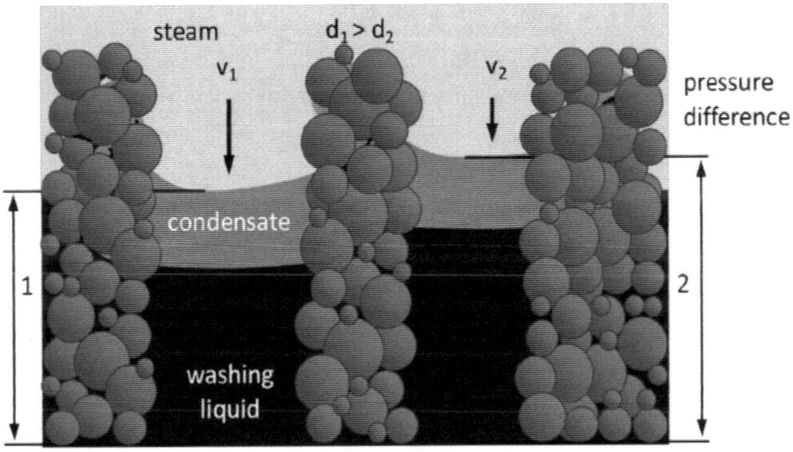

Figure 2. Condensation front in a prewashed filter cake.

$$\frac{m_{St}}{m_S} = \frac{T_{\max} - T_{\min}}{\Delta h_v}\left(c_{p,S} + \frac{RM}{1-RM} \cdot c_{p,L}\right) \qquad 8$$

with

$$RM = \frac{m_L}{m_L + m_S} \qquad 9$$

After the condensation front travelled through the filter cake the steam breakthrough occurs. The residual moisture in the filter cake is mainly condensed steam, which can be reduced by applying further steam or dry air. An application of air on the filter cake leads to a large drop of residual moisture as no further condensation occurs but the heated residual moisture evaporates in the dry air stream and is dragged away. This effect attenuates with time. A longer application with unsaturated steam then leads to lower residual moisture content [19], but higher costs as well. Maximal benefit of steam pressure filtration is given when the costs of steam production do not exceed the costs of conventional drying and pressurized air is used for further drying instead of steam after steam breakthrough.

Process Characteristics and Results

Tests in suspension times showed no significant increase in the bitumen recovery or decrease in the filtration resistance, which is indicator for mass through put, with a longer suspending. Another factor influencing the duration of filtration and the bitumen recovery is the oil sand/bitumen or bitumen/solvent ratio. The solids volume concentration depends on this ratio (see equation 2). If the ratio is too high, the bitumen is not solved entirely, the viscosity rises. A proper mechanical solid liquid separation is not possible. The optimal ratio of toluene and n-heptane in the suspending liquid is 30/70. At this composition fines affect filtration resistance least. The fines content of the later processed oil sand does not exceed 10 %. The optimal ratio of solvent to bitumen to be solved for oil sands of medium and poor quality is between 8 and 14, to gain a maximum bitumen recovery at the best process conditions. Thus the viscosity of the solvent with the diluted bitumen is small enough, to hinder a rising of the pressure loss over the filter cake height. If the ratio drops below 4 besides a higher pressure loss the formation of a covering layer especially for rich oil sands can occur (see figure 3). The covering layer consists of solids with a higher fines content and bitumen. The bitumen content in this layer is four times the bitumen content in the lower layer over the filter medium, where bitumen is enriched. Formation of the covering layer starts with the beginning of the filtration as the coarser particles tend to sediment fast. The finer parts with bitumen sink on the formed filter cake, continuously increasing filter cake resistance. In the application this phenomenon will not occur as blending of the processed oil sands guarantees a bitumen content of 10 %.

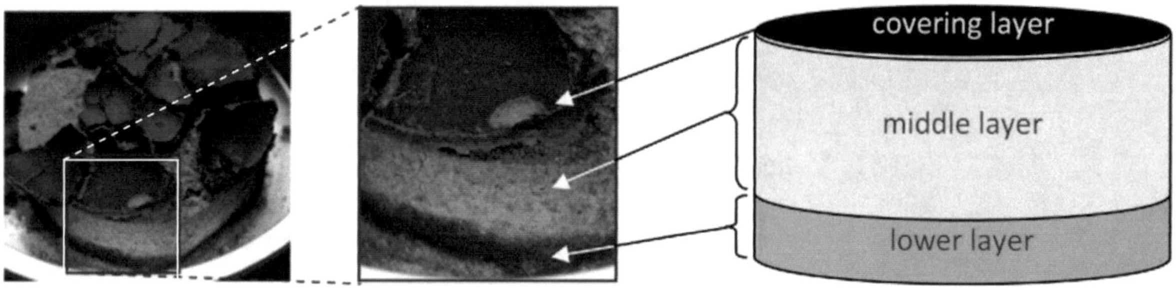

Figure 3. Layer formation during filter cake formation; dried covering layer, middle layer and lower layer.

In table 1 it is shown that a bitumen content above 12.8 % affects the filtration resistance r_c significantly. In processing this means throughput losses. The low and medium oil sand show larger differences in the bitumen content but the gained filtration resistances are stable. For rich oil sand a c_V of 0.3 is the upper limit at which a filtration resistance with a maximum of $3 \cdot 10^{14}$ m^{-2} occurs. The values in the table show maximum resistances at a good filterability. They are presupposed by the tendency of the formation of a covering layer. Variations of the resistances are not only driven by the process parameters but also by varying material parameters (natural source).

Table 1. Filtration resistances of different oil sand qualities at Δp=1 bar, c_V=0.23…0.33.

Oil Sand Quality	**Bitumen Content in %**	**lower limit r_c in m^{-2}**	**upper limit r_c in m^{-2}**

low	5	$5 \cdot 10^{11}$	$2 \cdot 10^{13}$
medium	12.8	$6 \cdot 10^{11}$	$2 \cdot 10^{13}$
rich	15	$2 \cdot 10^{13}$	$1 \cdot 10^{15}$

In the washing the applied toluene contents affect the further bitumen recovery. Least variations in the reached bitumen recovery are gained with a washing liquid with 50 % by weight of toluene. The bitumen recovery at this point of the process exceeds 90 % with a fully saturated filter cake after washing. Special phenomenon in washing is a concentration ratio exceeding one for low washing ratios. Reason for this is bitumen enrichment above the filter medium and in the outlet cone below the filter cake after filter cake formation. The uneven distribution of the bitumen over the filter cake height can affect the processing success negatively. Another special occurrence is a drop in bitumen concentration below one at washing ratios smaller one, which is typical for fingering and additionally to this for the uneven bitumen distribution described earlier.

The progress in the steam application can be described according to the washing with a washing curve, even if no washing effects occur during this last step. The washing curve can be summed up with the calculated condensation mass (see figure 4). The displacement ratio of the condensation front can be compared with the washing ratio (see equation 12 and 13).

$$W = W_W + W_{SPF} \quad (10)$$

with

$$W_{SPF} = \frac{\frac{m_{St}}{m_S} \cdot \frac{m_S}{\rho_{H2O}}}{\varepsilon \cdot h_K \cdot A} \quad (11)$$

The washing curve shows that for a medium oil sand on which steam is applied at 2 bar the concentration ratio decreases. As bitumen is not soluble in the condensation layer and the bitumen concentration within in the filter cake at the beginning of the steam appliance is very low, the bitumen recovery can only be increased to a very little extend in this last step. Recovery rates are raised around 2 points of percentage. The steam pressure chosen affects the bitumen recovery as well as the residual moisture. Tests show better bitumen recovery rates at lower steam pressures applied. Residual moisture is lower as well.

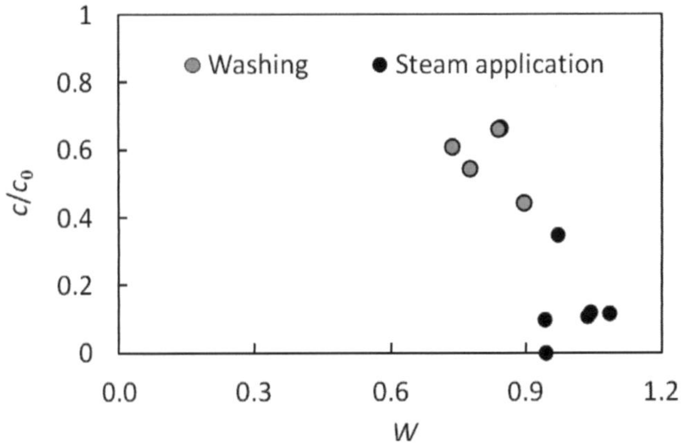

Figure 4. Washing curve for medium oil sand, solvent washed and steam applied at Δp=2 bar.

The specific steam input lies within a typical range of 50 to 250 kg per tones solid [20]. By the use of the sensible heat, brought into the system by the steam, the residual moisture can be decreased significantly with the application of dry air or superheated steam.

Acknowledgements

This work was possible with the cooperation with Siemens AG, Industry Sector and realized at the Institute of Mechanical Engineering and Mineral Processing. I want to thank all concerned parties.

Symbols

A	m^2	filtration area
c	$g\,l^{-1}$	concentration
c_p	$kJ\,kg^{-1}\,K^{-1}$	specific heat capacity
Δh_v	$kJ\,kg^{-1}$	enthalpy of evaporation
ε	-	porosity
η	$kg\,m^{-1}\,s^{-1}$	dynamic viscosity
k	m^2	permeability coefficient
m	kg	mass
p	bar	pressure
r_c	m^{-2}	specific filter cake resistance
RM	$kg\,kg^{-1}$	residual moisture
R_M	m^{-1}	specific filter medium resistance
S	-	saturation of filter cake
S_V	$m^2 m^{-3}$	specific particle surface
t	s	time
T	K, °C	temperature
V	m^3	volume
W	-	washing ratio
x	m	pore diameter

Greek Letters

ε	-	porosity
η	kg m^{-1} s^{-1}	dynamic viscosity
κ		concentration parameter

Indices

0	start-...; t=0
Bit	bitumen
F	filtrate
FC	filter cake
L	liquid
m	mass related
max	maximum
min	minimum
P	pore liquid
S	solid
Sol	solvent
SPF	steam pressure filtration
St	steam
V	solids volume
W	washing liquid

References

1. *Talk about Oil Sands*. 2013, Alberta Government. p. 2.
2. *Tar Sands Oil Mobile - FAQs*, in *A Community Coalition*. 2013, Tar Sands Oil Mobile: Alberta, Canada.
3. Schröder, B. *Der kanadische Ölsand-Komplex*. 2008 [cited 2008 02.08.]; Available from: http://www.heise.de/tp/artikel/27/27579/1.html.
4. Masliyah, J., et al., *Understanding Water-Based Bitumen Extraction from Athabasca Oil Sands*. The Canadian Journal of Chemical Engineering, 2004. **82**(4): p. 628-654.
5. CAAP. *Oil Sands Today*. Tailings Ponds 2013 [cited 2013 30.10.2013]; Available from: http://www.oilsandstoday.ca/topics/Tailings/Pages/default.aspx.
6. *Oil Sands Information Portal*. 2013 [cited 2013 27.11.2013]; Available from: http://environment.alberta.ca/apps/osip/.
7. Avid, B., et al., *Characterization of asphaltenes from Brazilian vacuum residue using heptane-Toluene mixtures*. Energy and Fuels, 2004. **18**(6): p. 1792-1797.
8. George, S., *Measurement and Modeling of Asphaltene-Rich Phase Composition*. 2009, University of Calgary.
9. Saadatmand, M., H.W. Yarranton, and K. Moran, *Rag Layers in Oil Sand Froths*. Industrial &

Engineering Chemistry Research, 2008. **47**(22): p. 8828-8839.

10. Speight, J.G., *The chemistry and technology of petroleum*. 2010: CRC press.

11. Hupka, J., J.D. Miller, and J. Drelich, *Water-Based Bitumen Recovery from Diluent-Conditioned Oil Sands*. The Canadian Journal of Chemical Engineering, 2004. **82**(5): p. 978-985.

12. Woods, J.R., et al. *Comparison of bitumens from oil sands with different recovery profiles*. 2009. Montreal, QC.

13. Czubak, T., *Untersuchung von Bitumina zur Herstellung von Öl-in-Wasser-Emulsionen*, in *Fakultät für Mathematik, Informatik und Naturwissenschaften, Lehrstuhl für Makromolekulare Materialien und Oberflächen (Physikalische Chemie) [153510]*. 2012, RWTH Aachen.

14. Wakeman, R. and E.S. Tarleton, *Fluid Dynamic Background to filtration processes*, in *Solid Liquid Separation: Principles of Industrial Filtration*. 2005.

15. Louhi-Kultanen, M., et al., *Filter cake washing: Partial dissolution of organic particles and real-time monitoring based on Raman spectroscopy*. Separation and Purification Technology, 2008. **59**(3): p. 270-276.

16. Heuser, J., *Filterkuchenwaschprozesse unter besonderer Berücksichtigung physikalisch-chemischer Einflüsse*, in *Universität Karlsruhe*. 2003, Shaker: Aachen.

17. Ruslim, F., et al., *Filter cake washing for cleaning soluble crystalline solid particles*. Filterkuchenwäsche zur aufreinigung löslicher kristalliner feststoffpartikel, 2006. **78**(6): p. 715-719.

18. Wakeman, R., *Filter cake washing*, in *Solid±liquid separation*. 1990.

19. Peuker, U.A. and W. Stahl, *Steam pressure filtration: Mechanical-thermal dewatering process*. Drying Technology, 2001. **19**(5): p. 807-848.

20. Wilkens, M. and U.A. Peuker, *Grundlagen und aktuelle Entwicklungen der Filterkuchenwaschung*. Chemie Ingenieur Technik, 2012. **84**(11): p. 1873-1884.

Optimal Control of Crystal Size Distribution for Nonisothermal Antisolvent Crystallization

Navid Ghadipasha , Jose Romagnoli

Department of Chemical Engineering, Louisiana State University, Baton Rouge, LA

American Institute of Chemical Engineers
April 2014

PSE @ LSU LSU

PROCESS SYSTEMS ENGINEERING AT LOUISIANA STATE UNIVERSITY

Outline

- Introduction
- Image-based technique
- Crystallization modeling
- **Control design and implementation**
- Conclusion

2

Crystallization

- A widely used technique in pharmaceutical and food industries
- Powerful separation process
  - High purity product
  - Size and shape control

LSU

3

Mechanism and Methods

- The driving force in crystal formation is supersaturation: $C - C^*$
- Can be achieved by several methods:
  - Cooling
  - Evaporation
  - Anti-solvent addition

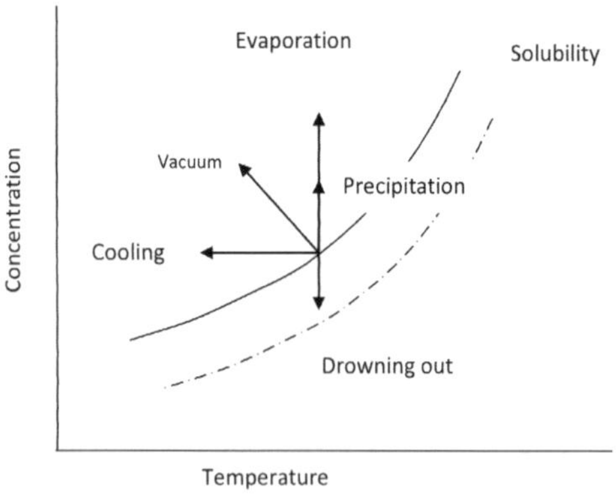

LSU

4

Crystal Size Distribution

- Affects down stream processes
- Can be related to supersaturation

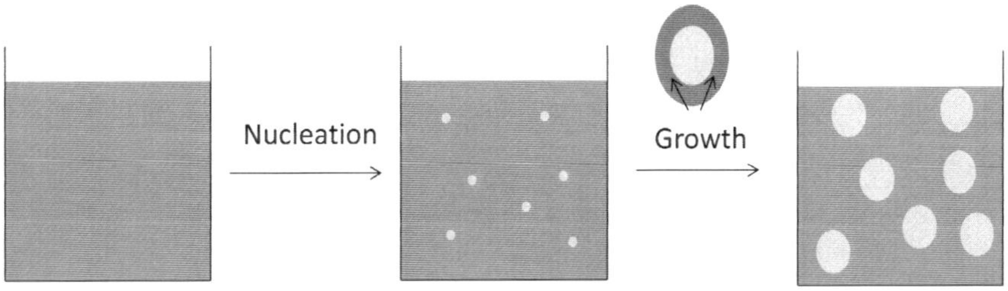

5

Research Objective

- Developing a ***model-based optimal strategies*** for cooling anti-solvent crystallization operations targeting crystal size distribution

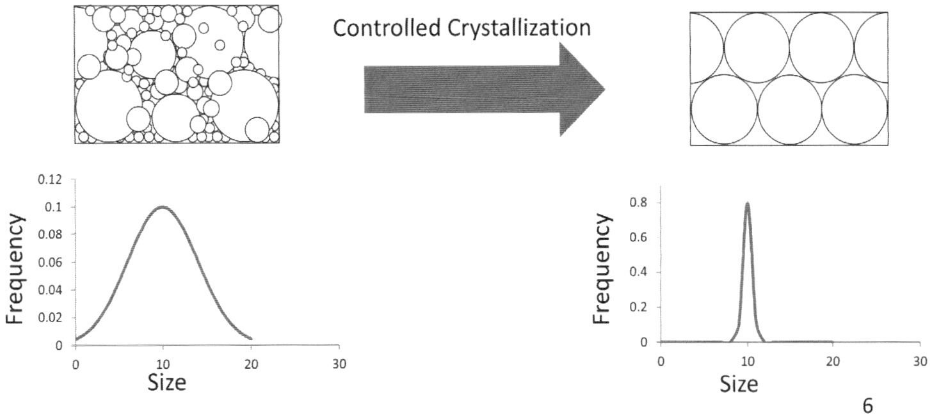

6

Control Basic

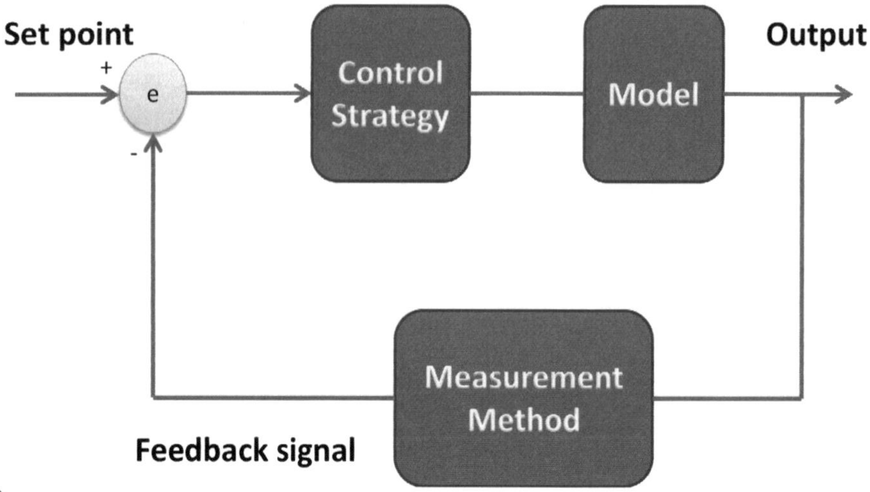

LSU

7

Image-based technique

- Image provides a close view of the crystallization evolution by estimating crystals characteristics like shape, CSD
- Developments in high speed imaging devices and Powerful computers at reasonable costs
- Determination of crystal mean size can be achieved by automatic methods (on-line process control and monitoring)

LSU

8

Experimental Setup

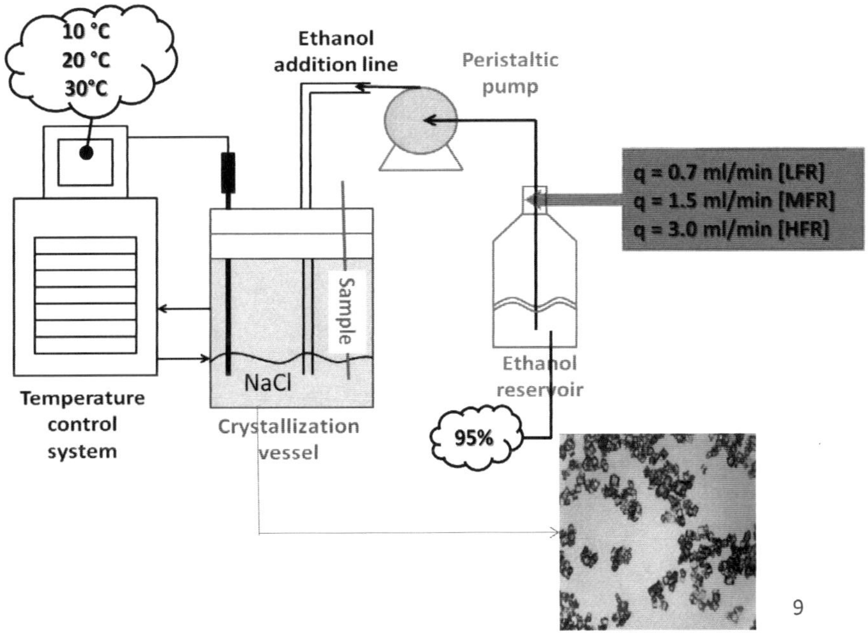

LSU

9

Fractal Dimension

- Fractal Dimension is selected as texture feature (1<FD<2).

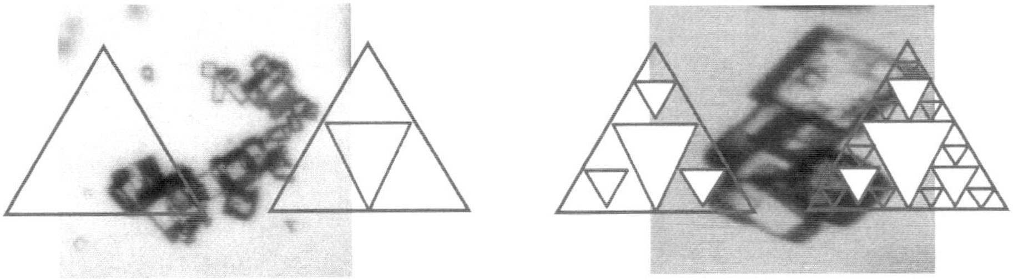

- Irrespective of whether crystals are individual or cluster
- Obtained by Fractal-Wavelet algorithm

LSU

10

Artificial Neural Network

- Strong ability in capturing complex input/output relationships
- Training and testing network by standard data.

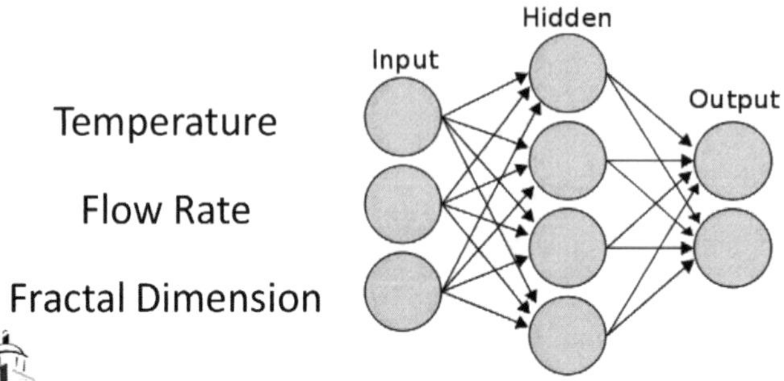

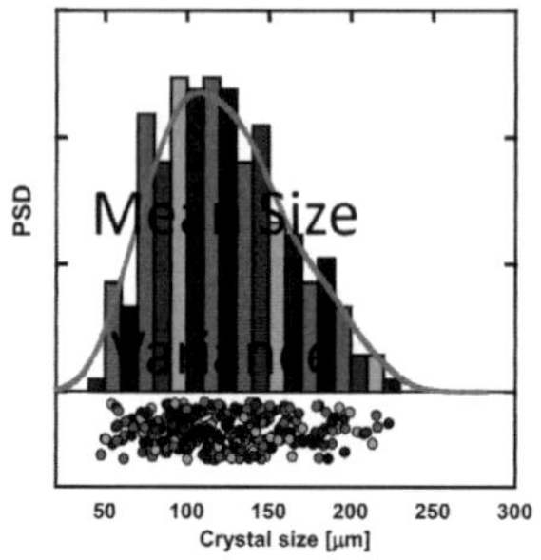

11

Control Basic

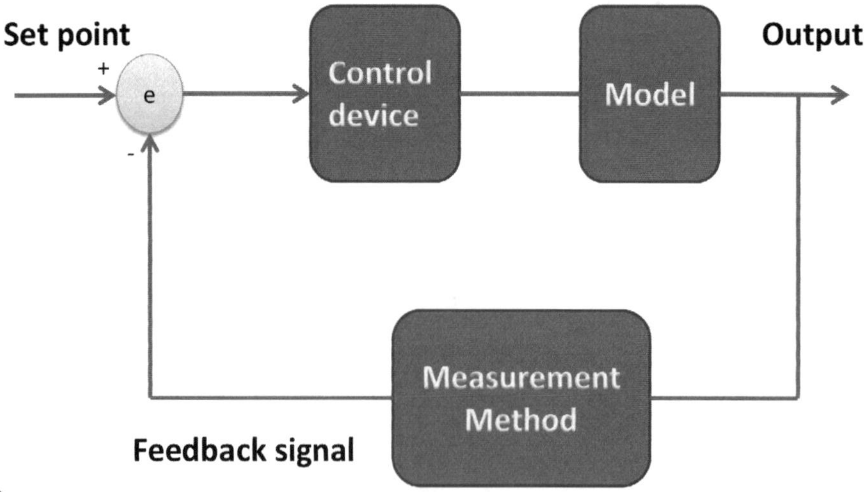

12

Modelling

Population Balances:

- Nonlinear PDE
- Describe different phenomena, such as the crystal growth
- analytical solution only considering some assumptions and neglecting some phenomena

Stochastic Models:

- Using Fokker-Planck Equation (FPE) in order to obtain the pdf of the system
- FPE is a linear PDE and, when the coefficients are linear too, it is possible to find an analytical solution

LSU

13

Fokker-Planck Equation (FPE)

$$\frac{\partial \psi(y,t)}{\partial t} = D\frac{\partial^2 \psi(y,t)}{\partial y^2} - \frac{\partial}{\partial y}(h(y,t;\theta)\psi(y,t))$$

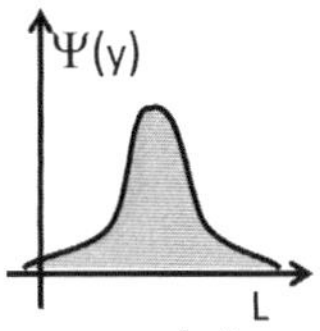

$y = lnL$

Diffusivity Term

Drift Term

$$\psi(y,t_0) = \frac{1}{\sigma_{y0}\sqrt{2\pi}}\exp\left[-\frac{(y-\mu_{y0})^2}{2\sigma_{y0}^2}\right]$$

IC: Normal Distribution

$$\frac{\partial \psi(+\infty,t)}{\partial y}=0$$

$$\frac{\partial \psi(-\infty,t)}{\partial y}=0$$

Boundary conditions

Gompertz Model (GZ-ML)

Linear PDE with linear coefficients

$$\frac{\partial \psi(y,t)}{\partial t} = D\frac{\partial^2 \psi(y,t)}{\partial y^2} - \frac{\partial}{\partial y}\left[r\left(1-\frac{y}{K}\right)\psi(y,t)\right]$$

LSU

14

Analytical Solution

$$\frac{\partial \psi(y,t)}{\partial t} = D\frac{\partial^2 \psi(y,t)}{\partial y^2} - \frac{\partial}{\partial y}\left[r\left(1-\frac{y}{K}\right)\psi(y,t)\right]$$

$$\psi(y,t) = \frac{1}{\sigma_y(t)\sqrt{2\pi}}\exp\left[-\frac{(y-\mu_y(t))^2}{2\sigma_y{}^2(t)}\right]$$

Gaussian Distribution

Mean Size Evolution

$$\mu_y(t) = \int_{-\infty}^{+\infty} y\Psi(y,t)dy = K\left\{1-\left(1-\frac{\mu_{y0}}{K}\right)\exp\left[-\frac{r(t-t_0)}{K}\right]\right\}$$

Variance Evolution

$$\sigma_y{}^2(t) = \int_{-\infty}^{+\infty}(y-\mu_y(t))^2\Psi(y,t)dy = \sigma_{y0}{}^2\exp\left[-\frac{2r(t-t_0)}{K}\right] + \frac{DK}{r}\left\{1-\exp\left[-\frac{2r(t-t_0)}{K}\right]\right\}$$

LSU

15

Global Model

Experimental Runs:

9 combinations of q e T

27 FPE parameters

0.7 ml/min	10 °C	*r* (growth velocity)
1.5 ml/min	20 °C	*K* (asymptotic dimension of crystals)
3.0 ml/min	30 °C	*D* (FPE Diffusivity, “dispersion”)

$$r_i(q,T) = \gamma_{i,0r} + \gamma_{i,1r}q + \gamma_{i,2r}T$$
$$K_i(q,T) = \gamma_{i,0K} + \gamma_{i,1K}q^2 + \gamma_{i,2K}T^2$$
$$D_i(q,T) = \gamma_{i,0D} + \gamma_{i,1D}q^2 + \gamma_{i,2D}T$$

LSU

16

Parameter Relationships

✓ The stochastic modeling is an alternative way of the more common population balances;
✓ Analytical solution;
✓ Antisolvent and temperature influence.

Increase	Asymptotic Mean Size (K)	Dispersion (D)	Growth Velocity (r)
q	↓	↑	↑
T	↑	↓	↓

LSU

17

Control Basic

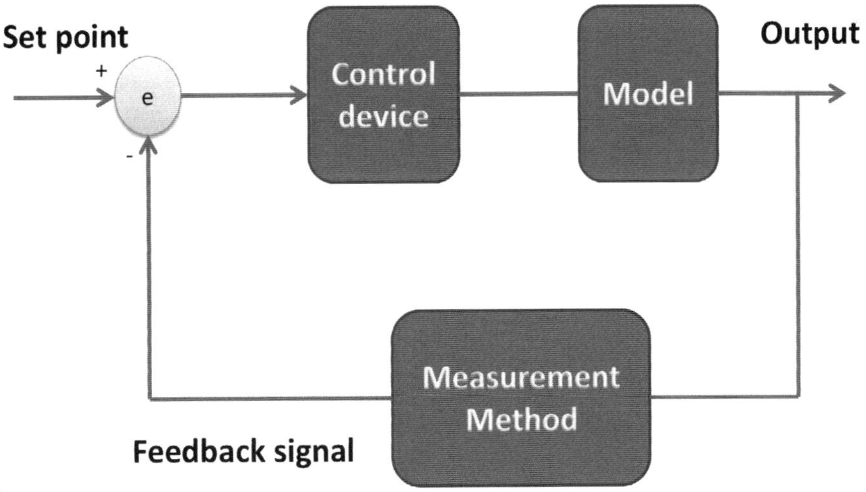

LSU

18

Control Challenges

- Occurrence of input multiplicities
- Ill-conditioning of the input-output relationship
- Semi-batch conduction mode (Time varying characteristic)

LSU

19

Input Multiplicities

- Same values of controlled variables for more than one set of manipulated variables
- Process ramification might happen
- Multiplicity detection by solving equations at asymptotic condition

$$\mu_{y,asymp} = K(q,T)$$ Asymptotic Mean Size

$$\sigma^2{}_{y,asymp} = \frac{K(q,T)D\,(q,T)}{r\,(q,T)}$$ Asymptotic Variance

LSU

20

Input Multiplicities

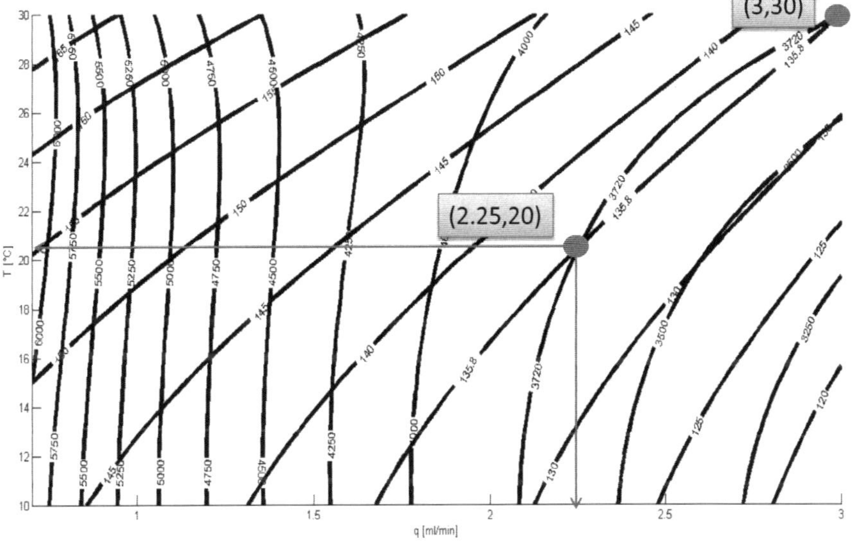

LSU

21

Ill Conditioning

- Sensitivity to model errors and possible difficulties for achieving acceptable control performance
- No direct guidance on the loop pairing between the manipulated and the controlled variables
- Gain matrix singularity

$$K_p = \begin{pmatrix} K_{p11} & K_{p12} \\ K_{p21} & K_{p22} \end{pmatrix} = \begin{pmatrix} \dfrac{\partial \mu_y}{\partial q} & \dfrac{\partial \mu_y}{\partial T} \\ \dfrac{\partial \sigma_y^2}{\partial q} & \dfrac{\partial \sigma_y^2}{\partial T} \end{pmatrix}$$

$$|K_{p,asymp}| = 0$$

LSU

22

Ill Conditioning

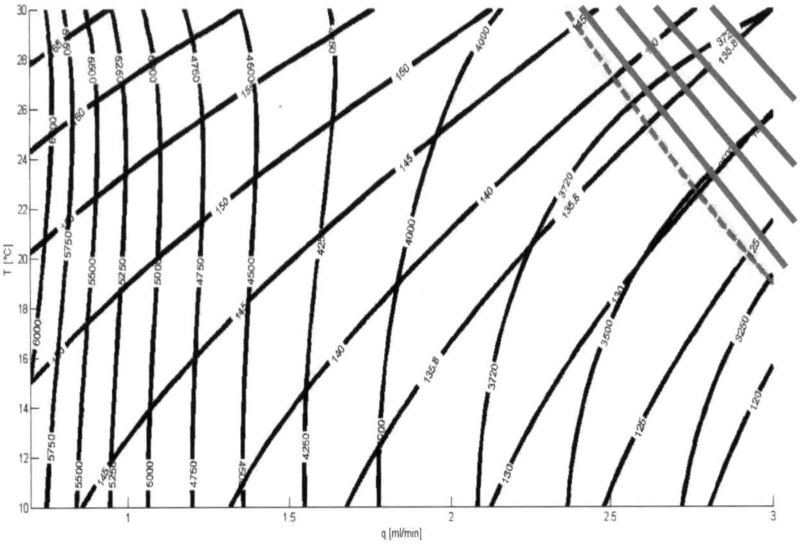

Two Stage controller

- Combination of an on-off controller and a feedback controller
- Feed-forward control as a result of static optimization procedure bring the system closed to the desired conditions

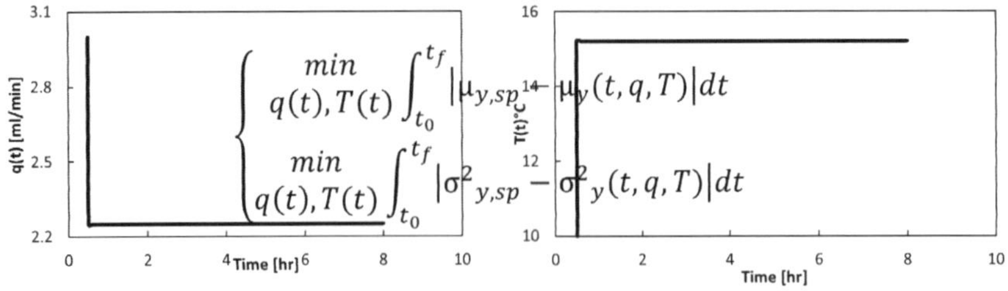

$$\begin{cases} \min\limits_{q(t),T(t)} \int_{t_0}^{t_f} \left| \mu_{y,sp} - \mu_y(t,q,T) \right| dt \\ \min\limits_{q(t),T(t)} \int_{t_0}^{t_f} \left| \sigma^2_{y,sp} - \sigma^2_y(t,q,T) \right| dt \end{cases}$$

- SISO control, based on IMC algorithm

IMC Design

- linear approximation of the mean dynamics

$$\mu(t) = \delta\mu + \mu^*$$
$$\sigma^2(t) = \delta\sigma^2 + \sigma^{2*}$$

- System transfer function

$$G_p(s) = \frac{K_p(q^*,T^*)}{\tau_p(q^*,T^*)s + 1}$$

$$\tau_p(q,T) = \frac{K\,(q^*,T^*)}{r\,(q^*,T^*)}$$

$$K_p(q,T) = \tau(q^*,T^*)\left[\left(1-\frac{\mu}{K}\right)\frac{\partial r}{\partial q} + \frac{r\mu}{K^2}\frac{\partial K}{\partial q}\right]_*$$

25

Control Schematic

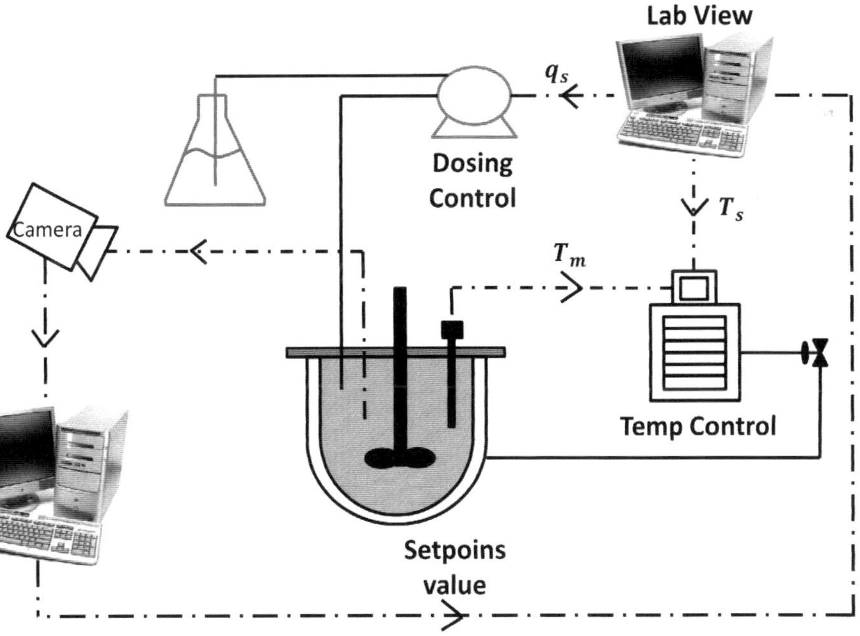

26

Control Results

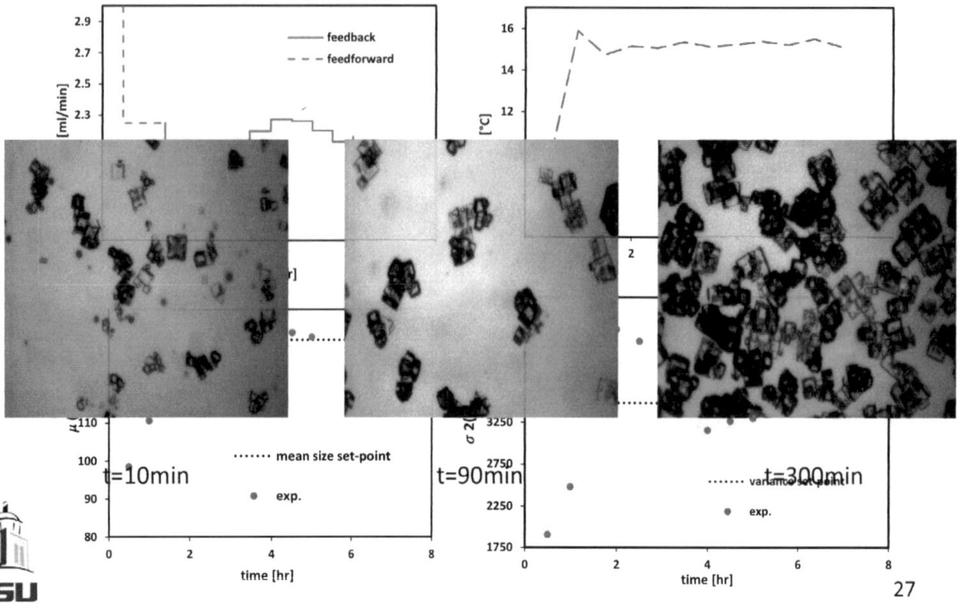

27

Conclusion

- Non-isothermal antisolvent strategy adds a second degree of freedom to be used to control the crystallization process
- Knowing exact transfer function of both, mean size of crystal and CSD variance, facilitates development of an internal mode-based controller
- System shows good controllability over mean size however controlling size variance is a demanding job

28

Thank you!

29

AUTHOR INDEX